Fashion Industry - An Itinerary Between Feelings and Technology

Edited by Riccardo Beltramo, Annalisa Romani and Paolo Cantore

Published in London, United Kingdom

IntechOpen

Supporting open minds since 2005

Fashion Industry - An Itinerary Between Feelings and Technology
http://dx.doi.org/10.5772/intechopen.78918
Edited by Riccardo Beltramo, Annalisa Romani and Paolo Cantore

Contributors
Amélia Maria Pinto Da Cunha Brandão, Mahesh Gadekar, Riccardo Beltramo, Jacopo Beltramo, Paolo Cantore, Annalisa Romani, Erica Varese, Anna Claudia Pellicelli, Kanchana Dissanayake, Dr. Kavita Chaudhary, Li Li, Jennifer Xiaopei Wu

First published in London, United Kingdom, 2020 by IntechOpen
IntechOpen is the global imprint of INTECHOPEN LIMITED, registered in England and Wales, registration number: 11086078, 7th floor, 10 Lower Thames Street, London, EC3R 6AF, United Kingdom
Printed in Croatia

British Library Cataloguing-in-Publication Data
A catalogue record for this book is available from the British Library

Additional hard and PDF copies can be obtained from orders@intechopen.com

Fashion Industry - An Itinerary Between Feelings and Technology
Edited by Riccardo Beltramo, Annalisa Romani and Paolo Cantore
p. cm.
Print ISBN 978-1-78984-119-0
Online ISBN 978-1-78984-120-6
eBook (PDF) ISBN 978-1-78985-259-2

For EU product safety concerns:
IN TECH d.o.o., Prolaz Marije Krucifikse Kozulić 3, 51000 Rijeka, Croatia, info@intechopen.com or visit our website at intechopen.com.

Meet the editors

Riccardo Beltramo is Full Professor at the University of Turin, Department of Management. He works as the Coordinator of the Commodity Sciences area. Since 1986 he has worked in the field of industrial ecology and carries out research on integrated management systems applied to manufacturing, service activities, and industrial areas. He currently teaches environmental management systems, tourism eco-management, and industrial ecology. He is President of the Bachelor Degree in Business Administration. Maker soul, in the field of Internet of Things has invented the Scatol8® System, a remote sensing system to monitor, show, and process environmental and management variables, giving life to the innovative startup "Lo Scatol8 per la Sostenibilità srl," an academic spinoff of the University of Turin.

Annalisa Romani is Professor of New Technology and Environmental Chemistry, Industrial Chemical Plants and Product Chemistry, Associate Professor of Commodity Sciences SECS-P/13, President and Scientific Referee of the Inter-Departmental Laboratory DiSIA-PHYTOLAB (Pharmaceutical, Cosmetic, Food Supplement Technology, and Analysis), Scientific and Technological Pole, University of Florence, Director of the Laboratory of Commodity and Quality of Resources, Faculty of Economy, University of Florence, Scientific Referee of PIN QuMAP Laboratory (Quality of Goods and Reliability of the Product), Prato, University of Florence, and Scientific Referent of Cluster C.H.I.Co, an area of pharma food. She is the author of 200+ publications, with more than 100 in international scientific journals. She is also the holder of nine national and European patents. She has an H-Index of 41.

Paolo Cantore is a computer science engineer and graduated at the Polytechnic of Turin. His work experiences mainly concern wireless sensor networks for environmental monitoring and management systems. He has participated at many research projects working for the universities of Turin and Prato, and he took a PhD in business and management. He is currently doing a research grant at the University of Turin.

Contents

Preface XI

Section 1
Fashion and Sustainability 1

Chapter 1 3
Sustainability Initiatives in the Fashion Industry
by Jennifer Xiaopei Wu and Li Li

Chapter 2 21
Does Mass Customization Enable Sustainability in the Fashion Industry?
by D.G.K. Dissanayake

Chapter 3 33
Exploration of Bamboo Fabrics with Natural Dyes for Sustainability
by Kavita Chaudhary

Section 2
Trends in Fashion Industry: A Technological Approach 45

Chapter 4 47
The Counterfeit Market and the Luxury Goods
by Amélia Maria Pinto da Cunha Brandão and Mahesh Gadekar

Chapter 5 59
The RFId Technology for Monitoring the Supply Chain and for Fighting against Counterfeiting: A Fashion Company Case Study
by Erica Varese and Anna Claudia Pellicelli

Chapter 6 73
How to Make Fabrics Talk Environment: The Scatol8 per la Sostenibilità Way
by Riccardo Beltramo, Jacopo Beltramo, Paolo Cantore and Annalisa Romani

Preface

We approached the editing of this book on the fashion industry with a clear vision, formalized in the invitation to contribute.

Fashion is a lot more than providing an answer to primary needs. It is a way of distinction, of proclaiming a unique taste, and/or expressing the belonging to a group. Sometimes to an exclusive group. Currently, the fashion industry is moving towards hyperspace, to a multidimensional world that is springing from the integration of smart textiles and wearable technologies.

It is far beyond aesthetics, though new properties of smart textiles have allowed designers to experiment with astonishing forms and expressions. There are new functionalities made possible by the interactions between wearable technologies, fabrics, and the wearer.

There are also surprising contrasts and challenges: a new life for natural fibers, fabrics, and dyeing techniques, environmental friendliness, rediscovered by eco-fashion, and "artificial apparel," made of microprocessors, sensors, and actuators. A movement begun by makers, startups, and microcompanies that today solicits the interest of famous fashion makers. Is the fashion industry taking profit from this revolution? How is this revolution affecting the strategies of the fashion industry?

With this spirit, we have selected scientific works capable of composing the multifaceted world of fashion in a single design.

Now, the outcome of the process is in your hands.

The theme of sustainability takes to the stage and becomes the protagonist of the first part of the book. Very complex issues, involving economic, social, and environmental aspects, are dealt with in two phases: initially, through three contributions on tools that allow companies to collect and provide evidence of their environmental motivation; therefore, through effective examples of the application of technologies and methodologies aimed at improving communication with consumers and eco-compatible solutions.

Chapter 1 discusses "Sustainability initiatives in the fashion industry," by Li Li, and proposes an itinerary in four steps: an overview of the most concerning environmental impacts caused by the fashion industry; current leading collective sustainability campaigns mobilizing the fashion industry; current available benchmarks and tools for measuring environmental impact of the textile lifecycle; and examples of how companies in the fashion industry are executing sustainability initiatives in their products or processes.

The goal of the second chapter, written by D.G.K. Dissanayake, is to provide a detailed and robust answer to the question that stands also as the title of the chapter: "Does mass customization enable sustainability in the fashion industry?" The starting point is the consideration of the dimension, in environmental form, of the evolution of the fashion industry, in particular due to the diffusion on fast fashion. The search for new business models seems to be the only way out. A promising consideration is the mass customization strategy and the author discusses the potential of this alternative, considering seven key elements that could possibly enhance sustainability. Innovation drives industry to ever cleaner and more efficient technologies. It's a matter of ethical management and of being able to provide evidence to clients and stakeholders.

Chapter 3, "Exploration of bamboo fabric with natural dyes for sustainability," by Kavita Chaudhary, shows the opportunity of bamboo viscose, a regenerated cellulosic fiber. The competitive advantages of bamboo as a provider of fiber, compared to other natural fibers, are described in terms of physical and chemical properties that contribute to the performance characteristics of fabrics. The chapter makes a comparative study of natural-dyed, printed bamboo, and cotton fabrics; it shows through a sequence of duly illustrated experiments the developing process of garments and natural-dyed bamboo fabric, thus providing an environmentally friendly and aesthetically challenging proposal.

The second part includes three chapters: two of them are focused on counterfeit. Pernicious behaviors, which are recorded with increasing concern, cause damage to consumers and to companies producing goods, in the parcel of luxury goods. Counterfeits are harmful both to the manufacturers of authentic products and to the buyers. The economic damage suffered by the company is direct, but it is only a small percentage of the reputational damage. The damage suffered by buyers is direct, when they believe they are buying a genuine product, and indirect for the problems that may be caused to their health when dangerous (and forbidden) substances have been used in the making process.

"The counterfeit market and the luxury goods," the fourth chapter, by Amélia Brandão, focuses on the counterfeit market, its influence on luxury consumption, and consumers' drivers for the counterfeit. The chapter also discusses innovative ways by which authenticity of luxury goods can be verified.

Erica Varese and Anna Claudia Pellicelli explore "The RFId technology for monitoring the supply chain and for fighting against counterfeiting. A fashion company case study" in Chapter 5. Their study involves an Italian company, Oscalito, which has adopted RFId technology as a valid support not only to monitor the supply chain, especially with reference to inventory management, waste disposal, logistics, and transport, but also to protect the Italian origin of production.

Chapter 6 contains a proposal sprung from our research team, joined by the academic spinoff company of the University of Torino, "Lo Scatol8 per la Sostenibilità srl." It describes the integration among devices inspired by the Internet of Things technologies and garments. The lifecycle thinking approach is applied to the evaluation of alternative fabric upcycling scenarios. In addition, wearable technologies make up a new dimension in the field of awareness raising. People, through the application of Scatol8's system, are allowed to witness their environmental consciousness, wearing interactive garments that extend the life of used products, even improving them. In the meantime, they contribute to the construction of maps of environmental quality, depending on the sensors that are hand stitched on the new garments.

We trust that you will be able to find interesting topics and ideas to further explore what you are most passionate about. We express our gratitude to all the authors with whom we have opened a collaborative and fruitful dialog, which we hope will meet with the favorable opinion of the readers.

Riccardo Beltramo
University of Torino,
Lo Scatol8 per la Sostenibilità srl,
Torino, Italy

Annalisa Romani
University of Firenze,
Lo Scatol8 per la Sostenibilità srl. Firenze, Italy

Paolo Cantore
University of Torino,
Torino, Italy

Section 1

Fashion and Sustainability

Chapter 1

Sustainability Initiatives in the Fashion Industry

Jennifer Xiaopei Wu and Li Li

Abstract

A heightened awareness toward the fashion industry's environmental impact has emerged in recent years, stirred by mounting evidence of intensified global clothing consumption and driven by the increased accessibility and affordability of clothing. In the last 3 years, the release of several comprehensive reports years detailing the extent of the fashion industry's environmental impact, as well as the founding of several fashion industry-targeted sustainability campaigns (e.g., the "2020 Commitment" of the Global Fashion Agenda), has not only helped draw a great deal of attention to the issues but has also triggered an evident wave of intention toward a concrete, quantifiable action. With the abundance of information surrounding the subject of sustainability in the fashion industry, this chapter intends to provide an overview of (1) the most concerning environmental impacts caused by the fashion industry, (2) current leading collective sustainability campaigns mobilizing the fashion industry, (3) current available benchmarks and tools for measuring environmental impact of the textile life cycle, and (4) examples of how companies in the fashion industry are executing sustainability initiatives in their products or processes. Finally, the chapter will conclude with some of the current challenges and future opportunities in sustainability confronting the fashion industry.

Keywords: fashion industry, Textiles and Apparel, sustainability, environmental impact, sustainability initiatives

1. Introduction

The taxing impact the fashion industry has had on the environment is by no means a new revelation—having accumulated a great deal of evidence over the years. However, unlike in the past when "sustainability" seemed more like an ideal adopted by individual, niche grassroot organizations, it is now considered a core value globally across the fashion industry. The fashion industry's recent wave of intentional action toward sustainability is in part motivated by several comprehensive and revealing industry sustainability reports released in the last 3 years [1–3], but moreover it is a collective response to the recent fashion industry-specific sustainability campaigns such as the "2020 Commitment," spearheaded in the last 2 years by several sustainability-driven coalitions (e.g., the Global Fashion Agenda and the Waste and Resources Action Programme UK), which have rallied formal commitments from a significant portion of the fashion industry toward concrete, quantifiable action for sustainability by 2020.

IntechOpen

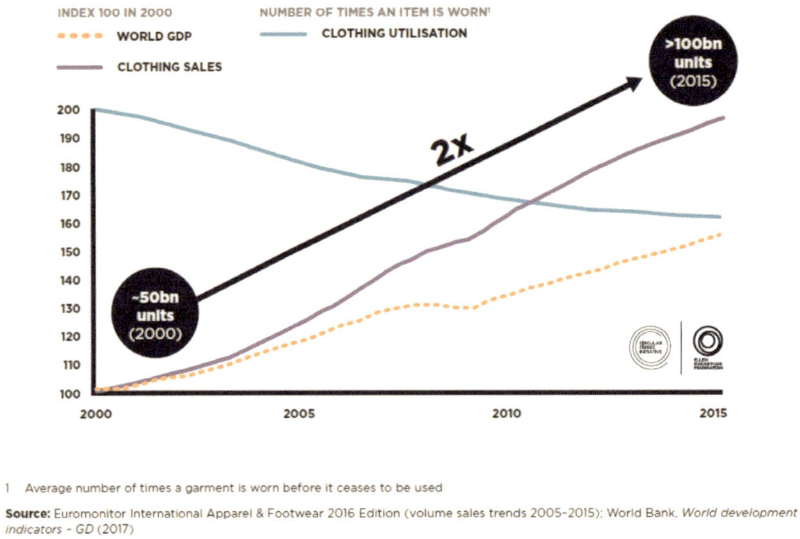

Figure 1.
Growth of clothing sales and decline in clothing utilization since 2000. Source: World bank, [4].

The heightened concern toward the fashion industry's environmental impact is also stirred by evidence of intensified global clothing consumption—which according to data from the World Bank [4] has doubled from around 50 billion units of clothing sales in 2000 to over 100 billion units in 2015 (see **Figure 1**). This dramatic increase in clothing consumption has been fueled by fast fashion, an increasingly bargain-driven consumer, increased accessibility via an expanding online shopping landscape, and more buying power from a growing middle class, especially in emerging economies such as China (projected to surpass the United States "as the largest fashion market in the world" in 2019, according to McKinsey FashionScope [5]). Unfortunately, the increased accessibility and affordability of clothing simultaneously propagated not only a culture of excessive consumption but also a quicker disposal of clothing, as exemplified by an approximately 20% decrease in the average number of times a garment is worn before it is abandoned as shown in **Figure 1**.

Given the abundance of information surrounding the subject of sustainability in the fashion industry from many sources, there is an opportunity for a collated overview on the subject. Therefore, the purpose of this article is to provide an overview of (1) the most concerning environmental impacts caused by the fashion industry, (2) current leading collective sustainability campaigns mobilizing the fashion industry, (3) current available benchmarks and tools for measuring environmental impact of the textile life cycle, and (4) examples of how companies in the fashion industry are executing sustainability initiatives in their products or processes. Finally, the article will conclude with some of the current challenges and future opportunities in sustainability confronting the fashion industry.

2. The environmental impact of the textile life cycle

In any given industry, each stage of the product life cycle poses an impact on the environment—by consuming environmental inputs (e.g., water for harvesting raw materials, fossil fuels to power manufacturing equipment, etc.) and releasing environmental outputs (e.g., carbon dioxide emissions from burning fossil fuels, landfill waste after product is disposed, etc.). For the fashion industry, the environmental

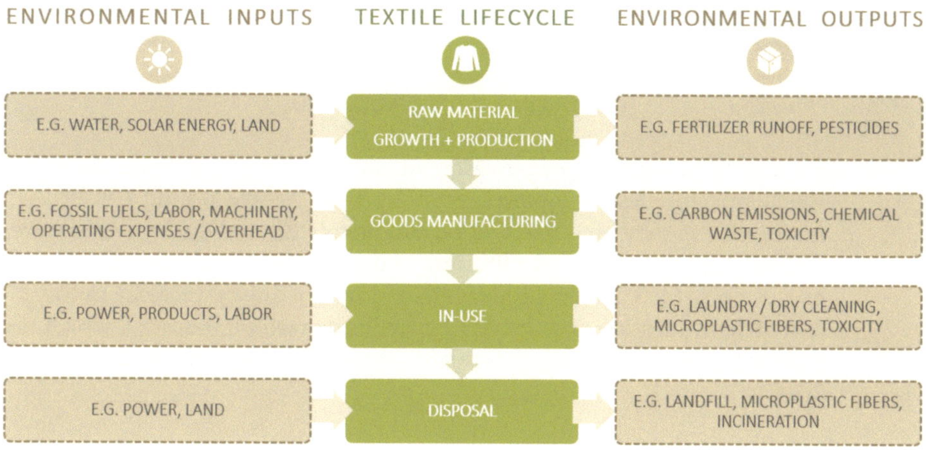

Figure 2.
Environmental impact (inputs and outputs) of the textile life cycle.

inputs and outputs of the textile product life cycle is reflected in **Figure 2**. (It is worthwhile to note that the term "life cycle" used is misleading in that the above chain of processes does not form a "cycle," but is instead linear sequence of events, with a definite beginning and end. A true cyclical life cycle would be indicative of recycling or reuse, feeding the end waste back into the system to be used again).

As shown, the inputs and outputs of the fashion industry's "textile product life cycle" pose impact on the environment, but it is the size of the impact which is staggering. This is partly due to the immense scale of the fashion industry, which has been evaluated to be a USD 1.3 trillion dollar industry [6], and the world's third largest manufacturing industry, after automotive and technology [7]. But also, according to a report by the Ellen MacArthur Foundation, data confirms that the greenhouse gas emissions produced by textile production exceeds that of international aviation and maritime shipping combined. If it continues down this path, it is projected that by 2050 it could account for 1/4 of the worlds' carbon emissions [1]. To put it into perspective further, the annual carbon footprint of the fashion industry's product life cycle (3.3 billion tons CO_2 emissions) is almost equivalent to that of 28 countries in the EU (3.5 billion tons) [7].

However, greenhouse gas emissions are not the only harmful environmental outputs from the fashion industry; it is just one of the numerous other inputs and outputs which have strenuous environmental implications, as exemplified in **Figure 2**. The below provides a summary, along with examples, highlighting some of the leading concerns (note that there are indeed many others; however, for the purpose of this condensed article, we will focus on the following):

- *Heavy consumption of depleting natural resources:*
    - For example, water consumption for cotton crops
    - For example, coal/natural gas (nonrenewable) energy to power manufacturing facilities
- *Polluting waste outputs (e.g., chemicals, pesticides, carbon emissions, etc.):*
    - For example, fertilizer/pesticide runoff from cotton crops

- Dyes/chemical waste from garment factories (e.g., for dyeing and washing processes)

- *Microplastic pollution (e.g., from synthetic fiber shedding):*

  - For example, shedding of polyester fibers (considered microplastics) in the laundry process: a domestic wash load can release around 700,000 fibers and, as they are unable to be completely filtered out by waste water treatment plants, end up infiltrating and accumulating in marine ecosystems [8]. This issue is exacerbated by the drastic increase in the annual consumption of polyester fibers in the fashion industry, which has grown exponentially, from 8.3 million tons in 2000, to 21.3 million tons in 2016 [6].

This section provided a condensed overview of the extent of the fashion industry's impact on the environment and highlighted the most concerning forms of impact. However, it is worth noting that the abundance of published data and literature on the environmental impact of the fashion industry is truly inundating and could easily extend beyond the scope of this section. The following section will present some of the current collective global sustainability campaigns which are striving to alleviate the environmental impact of the fashion industry in the future.

3. Collective global sustainability campaigns in the fashion industry

The intensified evidence of the fashion industry's impact on the environment in the last decade prompted the founding of several global sustainability campaigns within the last 3 years. These campaigns, spearheaded by sustainability-driven coalitions, are mobilizing companies across the fashion industry, collectively toward adopting sustainable materials and practices throughout their design, development, and supply chains, and have already garnered formal commitments from key players in the fashion industry which represent a sizable portion of the market. Two predominant global campaigns, initiated in 2018, are summarized below:

- ***The "2020 Circular Fashion System Commitment," introduced by the Global Fashion Agenda***

  - *Mission/action points:*

    The Global Fashion Agenda is a leadership forum engaging the fashion industry toward sustainability [9]. Its "2020 Circular Fashion System Commitment" is a call on the fashion industry to commit toward a "circular fashion system," by taking concrete action on one or more of the following points:

    1. Implementing design strategies for *cyclability*

    2. Increasing the volume of *used* garments and footwear *collected*

    3. Increasing the volume of *used* garments and footwear *resold*

    4. Increasing the share of garments and footwear made from *recycled post-consumer textile fibers*

- Industry commitment (as of May 2018):

    1. Ninety-four companies signed on (represents 12.5% of the global fashion market), including ASOS, H&M, Nike, Inditex, Kering, and Target.

- ***The "Sustainable Clothing Action Plan (SCAP) 2020 Commitment," introduced by the Waste and Resources Action Programme (WRAP)***

    - *Mission/action points:*

    The SCAP (spearheaded by WRAP) is a collaborative framework and voluntary commitment for organizations to deliver industry-led targets of a 15% *reduction* in *carbon*, *water*, and *waste* in the clothing industry by [10]:

    1. Reinventing how clothes are *designed* and *produced*

    2. Rethinking how we value clothing by *extending life* of clothes

    3. Redefining what is possible through *reuse* and *recycling*

- Industry commitment (as of March 2019):

- Eighty companies signed on (represents 58.5% of the UK's retail sales volume), including ASOS, Marks and Spencer, Ted Baker, and others.

The action points of both these campaigns show an emphasis on cyclability—not just of materials but also practices—and reshaping the product life cycle toward circularity [10] (see **Figure 3**). The number of companies committed to these campaigns so far is a promising sign that sustainability is gradually becoming an integral factor in the fashion industry. Aside from the global sustainability campaigns such as above, another industry resource supporting companies toward sustainability is the various benchmarks and tools developed to help the fashion industry gauge the environmental

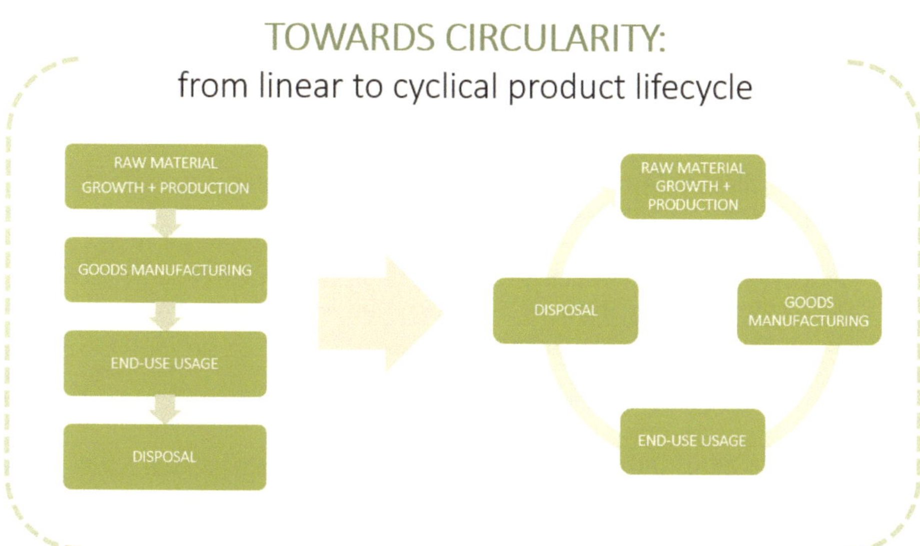

Figure 3.
A diagrammatic expression of the goal of "circularity" in the textile product life cycle.

impact of certain materials or processes and therefore help steer decisions accordingly. The following section will explore some of these tools and benchmarks.

4. Measuring environmental impact: benchmark and tools for the fashion industry

For companies in the fashion industry to become more cognizant and proactive about minimizing the environmental impact of their product life cycles, they would need to rely on a definitive benchmarks and tools to gauge the environmental impact of their decisions regarding product or processes. However, measuring environmental impact of such decisions can be very convoluted, as results tend to be conflicting depending on which angle it is viewed from. Here are some examples of the conflicting nature of environmental impact measures:

On the one hand, for example:

- A polyester shirt has more than double the carbon footprint of a cotton shirt (5.5 kg CO_2 emissions vs. 2.1 kg CO_2 emissions) [11].

But on the other hand:

- The processing for cotton produces a water footprint 20 times larger than that of polyester (see **Figure 4**).
- One kilogram of cotton—equivalent to the weight of a shirt and pair of jeans—can take as much as 10,000–20,000 liters of water to produce [10].
- For an organic cotton tote to make up for the environmental impact (water use, energy use, etc.) of a classic plastic bag, it would need to be used 20,000 times [12].

The following is an outline of three established benchmarks and tools, designed to enable the fashion industry (and other industries), to measure the environmental impact of certain decisions regarding their material use or processes employed:

- ***Higg Index, developed by the Sustainable Apparel Coalition:***

  It is described as "a suite of tools" that enables the measure and score of a company or product's "sustainability performance" at "every stage in their sustainability journey," aiming to provide a "holistic overview" that "empowers businesses to make meaningful improvements that protect the well-being of factory workers, local communities, and the environment" [13]. It encompasses the following tools:

  o *Product tools:*

  1. *Higg Materials Sustainability Index (MSI)*: "the apparel industry's most trusted tool to accurately measure the environmental sustainability impacts of materials," by scoring materials based on their environmental impact from fiber to fabric across five environmental impact parameters (global warming, water pollution, water scarcity, resource depletion, and chemicals) (see **Figure 5** for a sample screenshot of the Higg MSI interface)

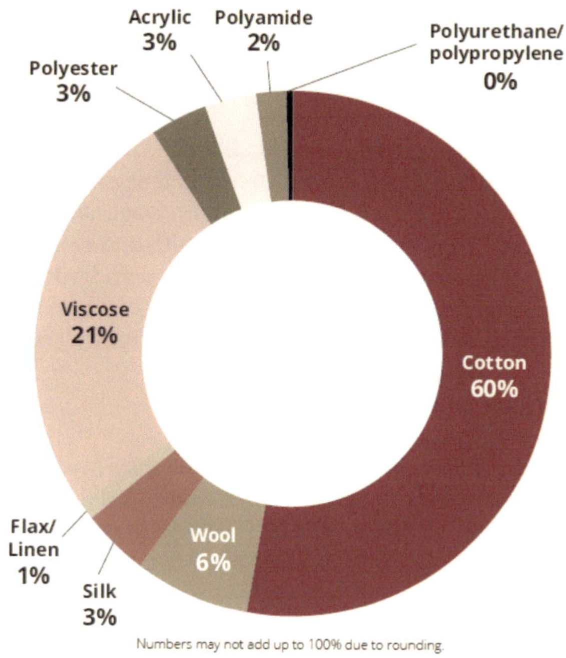

Figure 4.
Water footprint for the total processing phase of each fiber type for the UK (m3) in 2016. Source: Waste and resources action programme [14].

Figure 5.
Sample screenshots of the Higg Materials Sustainability Index (MSI) tool. Source: Sustainable apparel coalition [13].

2. *Higg Design and Development Module (DDM)*: "guides designers to combine their chosen materials for maximum positive impact, to select the most sustainable manufacturing techniques, and to consider the complete life-cycle of the product"

3. *Higg Product Module (PM)*: will measure the environmental impact a product (apparel, footwear, and textile products) generates throughout its life cycle when produced at industrial scale and be able to cross compare products with one another as well as which life cycle stages or production processes contribute the most impact (expected to launch in 2019)

- *Facility tools:*

1. *Higg Facility Environmental Module (FEM)*: measures the environmental impact of individual factories based on assessing factors such as

environmental management systems, energy use and greenhouse gas emissions, water use, wastewater, emissions to air (if applicable), waste management, and chemical use and management

2. *Higg Facility Social and Labor Module (FSLM)*: measures the social impact of individual factories based on assessing factors such as recruitment and hiring, working hours, wages and benefits, employee treatment, employee involvement, health and safety, termination, management systems, facility workforce standards and those of value chain partners, external engagement on social and labor issues with other facilities or organizations, and community engagement

o *Brand and retail tools:*

1. *Higg Brand and Retail Module (BRM)*: enables brands and retailers of all sizes to measure the environmental and social and labor impacts of their operations across a product's life cycle (from materials sourcing through its end of use). The environmental impacts measured include greenhouse gas (GHG) emissions, energy use, water use, water pollution, deforestation, hazardous chemicals, and animal welfare. The social and labor impacts measured include child labor, discrimination, forced labor, sexual harassment and gender-based violence in the workplace, non-compliance with minimum wage laws, bribery and corruption, working time, occupational health and safety, and responsible sourcing.

- ***MADE-BY Environmental Benchmark for Fibers, developed by MADE-BY in cooperation with Brown and Wilmanns Environmental, LLC:***

  It ranks 28th in the most commonly used fibers in the garment industry into 5 classes (Class A–E), based on the following measures: greenhouse gas emissions, human toxicity, eco-toxicity, energy, water, and land [15] (see **Figure 6**).

- ***Corporate Fiber and Materials Benchmark (CFMB) (formerly the Preferred Fiber and Materials Benchmark (PFMB)), launched by the Textile Exchange:***

  Launched in 2015, it is a leading industry-led, voluntary self-assessment tool which enables companies to systematically measure, manage, and integrate a preferred fiber and materials strategy into four key areas of mainstream business operations: corporate strategy, supply chain, consumption, and consumer engagement [16] (see **Figure** 7 for flowchart of this framework laid out). It also provides feedback on progress and performance in comparison to peers and the overall industry. As of 2018, 111 companies have partaken in the program (an increase of 106% since 2015).

As can be seen from the three examples above, there is a wide selection of benchmarks and tools for measuring environmental impact available to the fashion industry; however, there are some limitations to consider. For one, the wide selection can also be problematic as each of the different initiatives above accounts for slightly different factors or weighs them slightly differently; therefore the result obtained from one tool might not be consistent with that obtained from another. For example, based on the Higg Materials Sustainability Index, natural fibers like

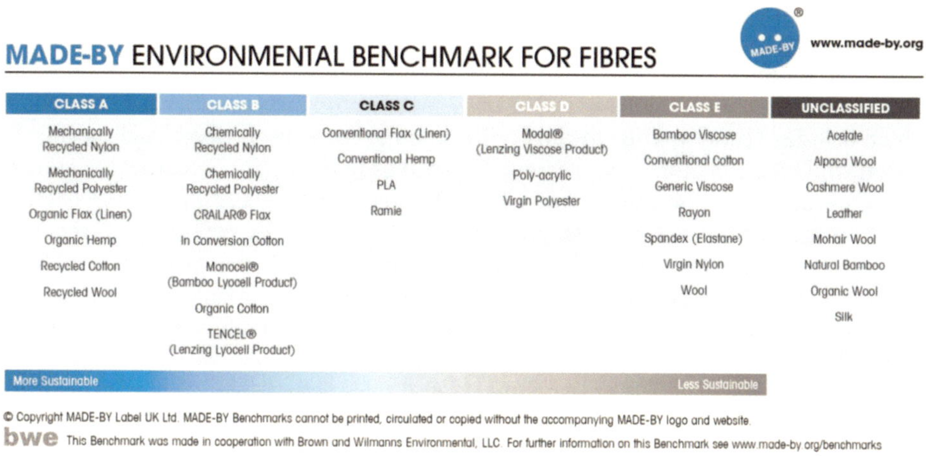

Figure 6.
The MADE-BY environmental benchmark fiber classification chart. Source: Common objective [15].

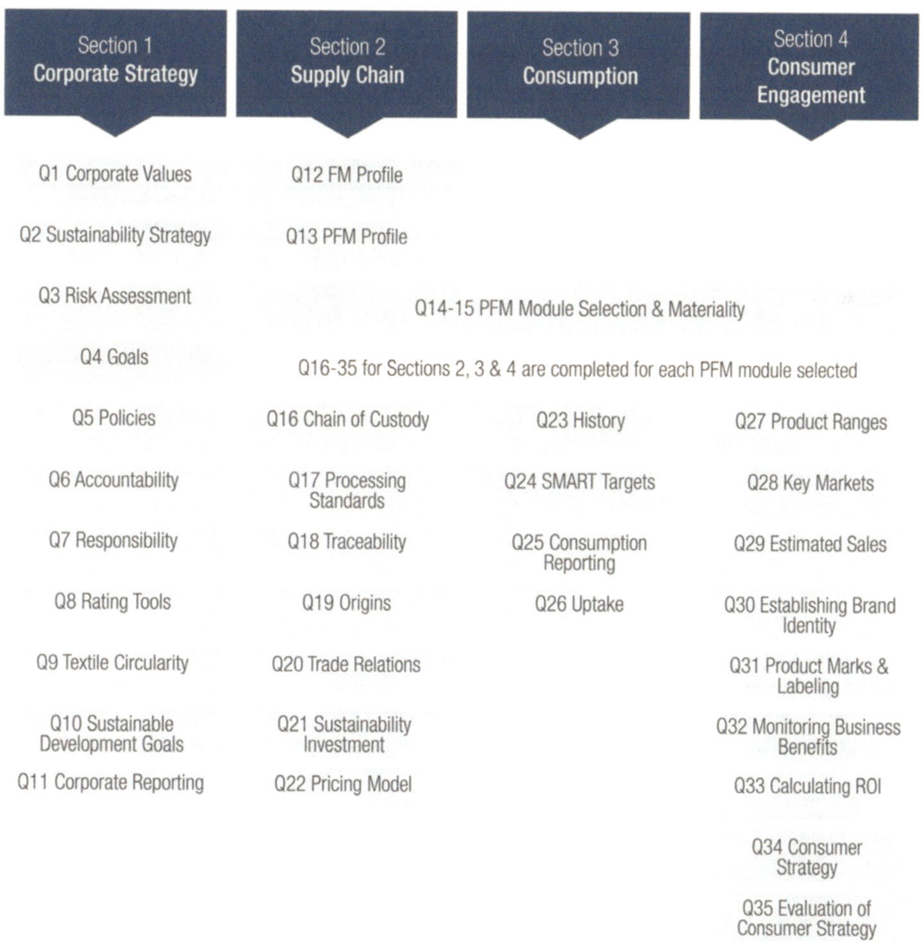

Figure 7.
Flowchart showing the framework of the textile exchange's corporate fiber and materials benchmark (CFMB). Source: Textile exchange, [16].

silk, cotton, and wool are assigned higher environmental impact scores (i.e., more damaging to environment) of 128, 98, and 82, respectively, while fossil-fuel-derived fibers like nylon, acrylic, and polyester have lower impact scores at 60, 52, and 44 [7]. This is because the Higg Index puts greater emphasis on fiber production, which is indeed more taxing on the environmental for natural fibers such as silk, cotton, and wool, as their procurement imposes a greater strain on natural resources (such as water, land, or animal welfare). Yet, in contrast, according to the MADE-BY Environmental Benchmark (**Figure 6**), fossil-fuel-based virgin nylon fibers and natural wool fibers are both ranked under the same Class E (the "least sustainable" category). Hence the availability of multiple benchmarks and tools could prove to be more incumbering than helpful when it comes to definitively measuring environmental impact.

Another limitation of these benchmarks and tools is that they do not sufficiently weigh in, or even overlook, the impact of the in-use phase of the textile product life cycle. The in-use phase here refers to the period when the textile product is being used for what it was made for. So, for a garment, that would mean the period from when it is purchased by a customer until it is no longer used or disposed of, which mostly involves its wearing and laundering. The research of Laitala et al. reveals that energy and water consumption during the laundering process varies greatly depending on fiber content of the garments [17]. Firstly, (see **Figure 8**) the research presents data which indicates that wool- and silk-based garments are 3–6 times more likely to be dry-cleaned than cotton- or synthetic-based garments and furthermore that dry cleaning uses 3–6 times (depending on the type of dry-cleaning

Main washing methods for clothing made of different materials

Washing Method by Fiber	Hand Wash	Machine Wash	Dry-Clean	Combination of Methods or Unknown
Cotton and cotton blends	8%	80.5%	6.5%	5%
Wool and wool blends	11%	35%	36%	18%
Synthetics and man-made materials	9.5%	71.5%	9%	10%
Silk	23%	25%	37%	15%

Estimated electricity usage of dry-cleaning and wet-cleaning processes/solvents

Cleaning Process/Solvent	Electricity Use (KWh/kg Textiles)
GreenEarth® (decamethylcyclo-pentasiloxane D5)	1.195
Hydrocarbon	0.783
LCO_2	0.681
PERC	0.586
Wet cleaning	0.205

Average wash load and temperature for different washing programs in Germany (n = 2763 wash cycles)

Washing Program	Number of Wash Cycles	Arithmetic Average Amount of Load with Standard Deviation (in kg per Wash Cycle)	Average Wash Temperature
Cotton	1967	3.4 ± 1.2	47.1
Synthetics	47	3.0 ± 1.0	44.1
Easy care	492	2.8 ± 1.3	38.8
Silk	1	-	20.0
Mix	74	3.7 ± 1.4	43.7
Wool	31	2.1 ± 1.1	30.3
Delicates	151	2.3 ± 1.2	34.6
Not specified	104	-	40.6

Figure 8.
Data on laundry requirements based on fiber content. Source: Laitala et al. [17].

process) more electricity than wet washing methods (which is the predominant laundering method for cotton- or synthetic-based garments). However, their research also shows that on average, the water temperature of the wash setting for cotton-based garments is about 17°C higher than that for wool-based garments. With polyester and other nonbiodegradable polymer fibers (e.g., acrylic and nylon), there is the developing concern regarding the shedding of fibers (microplastic) during the washing process which, being unable to be completely filtered out by standard waste water treatment plants, end up infiltrating and accumulating in marine ecosystems.

Another aspect which deserves more consideration by the benchmarks and tools is human ecology and not just environmental ecology. For example, there are man-made fibers derived from plants, such as polylactic acid (PLA) derived from corn, which are environmentally biodegradable, but not necessarily human biocompatible [18]. Therefore, the potential negative side effects or toxicity on human ecology is a factor which deserves equal attention in impact measures.

These limitations in the current benchmarks and tools are a clear reminder that measuring environmental impact of product or processes in the fashion industry is multifaceted and convoluted. Currently there is no prevailing, overriding benchmark or tool that provides a definite unanimous measure of environmental impact, so it is up to companies to adopt a holistic approach when developing a strategy toward sustainability.

5. Sustainability initiatives in the fashion industry

Having reviewed several sustainability campaigns and environmental impact measure benchmarks and tools relevant to the fashion industry today, this section will now proceed to provide insight into how companies and various players in the industry have responded, i.e., the kinds of strategic initiatives being taken toward sustainability. The sustainability initiatives will be categorized into two types: (1) front-end approach and (2) back-end approach.

5.1 Front-end approach

Within the context of this article, this refers to the integration of sustainable initiatives at the beginning stages (front-end) of the textile product life cycle, such as in the raw material sourcing and design and development processes. So, for example, a front-end sustainable initiative could be the decision to use "low environmental impact*" textile fibers as the raw materials for the textile goods being produced. A front-end sustainable initiative could also be manifested in the design and development process, for example, by utilizing digital tools to minimize the need for physical prototype samples or by training designers to adopt an eco-conscious mindset into their creations. (*Note that we are using the term "low environmental impact" textile fibers as opposed to "sustainable" or "eco-friendly" or "green" textile fibers because the latter terms can be misleading as there are no completely "sustainable/eco-friendly/green" fibers; all materials pose some impact. Furthermore, as discussed in the previous section, it is difficult to resolutely confirm the impact of a certain material, as there are many facets of environmental impact. Therefore "low environment impact" is a more accurate representation of what is possible to strive for in sustainable materials).

An industry example of a front-end approach to sustainability is the adoption of regenerated cellulosic fibers, such as Lyocell and Seacell, by various fashion companies particularly in lingerie and activewear [19]. With cotton, albeit a natural cellulosic fiber, bearing a hefty water footprint in the harvesting process, and with petrochemical-based synthetic fibers such as polyester and nylon bearing a hefty carbon footprint in the manufacturing process [20], regenerated cellulosic fibers can prove advantageous. They have the benefit of being biodegradable and derived from natural renewable resources (i.e., Lyocell is derived from wood pulp and Seacell is derived from seaweed) via a closed-loop manufacturing process, thereby consuming far less water and energy than traditional cotton, polyester, and nylon. Both Lyocell and Seacell also naturally carry antibacterial and fast-drying properties, which is why they are ideal for lingerie and activewear product.

A limitation of a front-end approach in tackling environmental impact is that it is still feeding more product in the fashion pipeline which will eventually end up at the end of the textile life cycle as waste by-product (even if it is biodegradable by-product) which needs to be managed accordingly. Therefore, in the following section, we will look at an approach which tackles the by-product end of the textile product life cycle.

5.2 Back-end approach

Within the context of this article, this is referring to sustainability initiatives which aim to minimize the environmental impact of the product and processes at the end of the textile product life cycle, e.g., at disposal. A prime example of this is exemplified in the now widespread initiatives of post-consumer textile recycling.

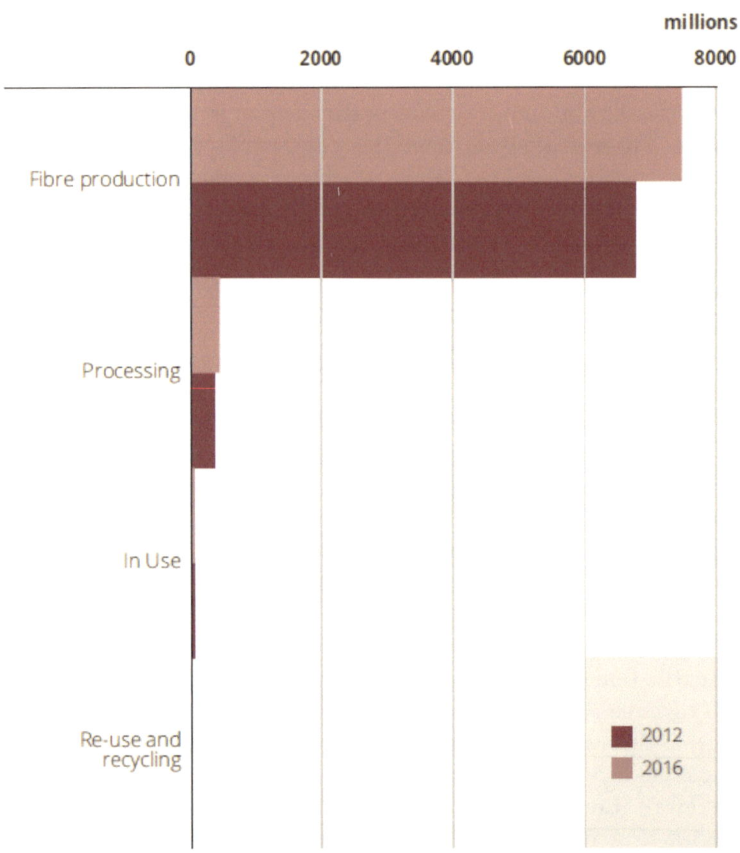

Figure 9.
Water footprint of clothing in the UK (m3) in 2012 and 2016, comparing life cycle stages. Source: Waste and resources action programme [14].

The reason recycled textiles have become so prevalent as a strategy to minimize environmental impact is not only because of the exponential supply of textile waste driven by intensified clothing consumption but more strategically because research has shown that the fiber production stage (extraction and processing) of the textile product life cycle has the greatest environmental impact in terms of water and carbon footprint, as shown in **Figures 9–10** [14]. Therefore, by recycling post-consumer textile waste back into the textile supply chain enables bypassing the heavy environmental toll of the fiber production stage.

There has been a great deal of research invested into textile recycling, from both the industry and academia. One notable advancement in textile recycling is exemplified by Garment-to-Garment (G2G) Recycle System, a closed-loop garment recycling retail concept supported by technology which enables the recycling of blended post-consumer garments, developed by HKRITA, in partnership with H&M and Novetex [21]. The Garment-to-Garment (G2G) Recycle System brings garment recycling to the retail level, therefore paving the way for garment recycling to be more accessible to the everyday consumer.

There are also several notable recycling initiatives which, instead of relying solely on post-consumer textile products, are derived from various kinds of post-consumer plastic waste. REPREVE is one example of this. Produced by the company Unifi, REPREVE is a brand of polyester fibers made from recycled post-consumer plastic waste (e.g., plastic bottles) [22]. The ability to convert various forms of

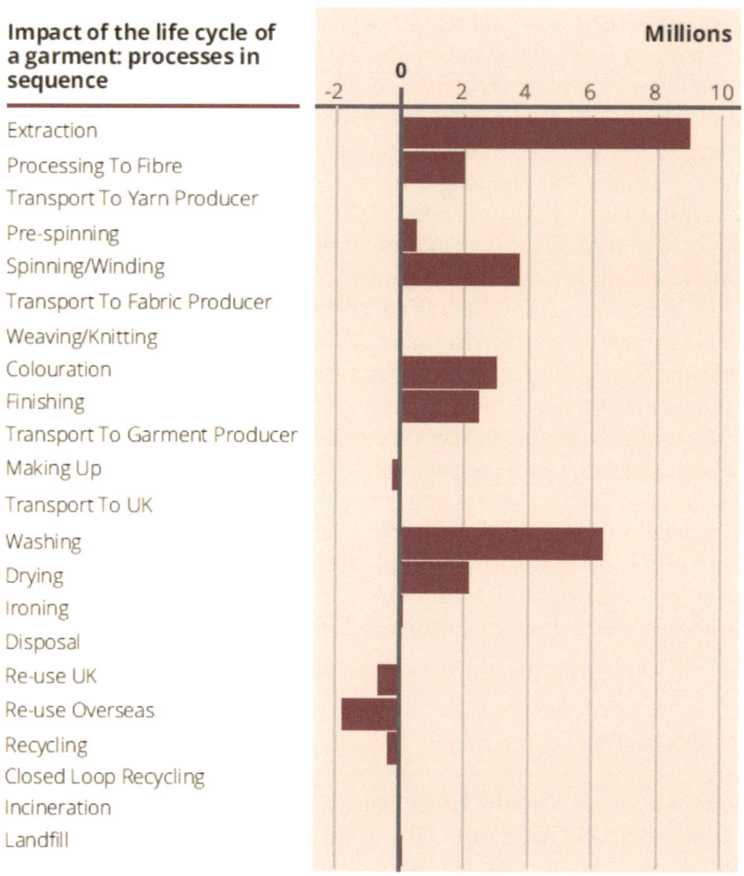

Figure 10.
Carbon footprint of clothing in the UK (t CO_2e) in 2016, by process. Source: Waste and resources action programme [14].

plastic waste into usable polyester textile fibers has the benefit of resourcefulness. Even though the conversion from post-consumer plastic waste to fiber requires energy and water input for the manufacturing process, according to Unifi, it is reportedly much less than that required for virgin polyester (energy consumption is 45% less, water consumption is almost 20% less, and over 30% less greenhouse gas emissions).

Over the past decade, there have been many encouraging advancements which have expanded back-end approach sustainability initiatives such as textile recycling. However, there remain limitations in the current textile recycling technologies. For example, due to the need for comprehensive shredding in breaking down post-consumer textile waste, the tensile strength of recycled cotton yarns is less than that of virgin cotton [23]. Furthermore, as recycled yarns are composed of a mixture of fibers which may have undergone different dyeing and finishing processes in their last life, even after cleaning and bleaching processes, they may not be able to achieve the same hand-feel and color vibrancy possible with virgin fibers, therefore limiting its design versatility. These are some examples of limitations which could be preventing a greater adoption of textile recycling in the industry.

6. Conclusion

This article has attempted to provide a current and overarching view on the most concerning environmental impacts of the fashion industry today, the leading global sustainability campaigns and benchmarks and tools established to help empower the fashion industry toward concrete action and, last but not least, examples of sustainability initiatives being implemented in the industry. The fashion industry's large-scale wave of movement toward sustainability is evident; however, there remain questions and challenges to be addressed, one being how successful the "2020 commitment" goals will be, with 2020 just around the corner, and considering how potentially disruptive any kind of change is in an industry which is built on long-established processes and practices and adheres to an inflexible, tight calendar. Furthermore, as discussed in this article, the array of benchmarks and tools available for measuring environmental impact can result in a convoluted process and conflicting, inconclusive information. Such challenges may deter a company from successfully achieving concrete changes toward sustainability.

Even if companies are able to navigate through the intricacies in evaluating environmental impact of a textile product or process, it is important to remember that the textile product life cycle is never impact-free (at least not in the foreseeable future), as it relies on the environment to provide various inputs and outputs. With this reality in mind, companies may find that making small but carefully holistically considered steps in the right direction can be much more effective than larger uninformed leaps when it comes to sustainability.

Acknowledgements

This work was funded by the Hong Kong Research Institute of Textiles and Apparel Limited (HKRITA), grant ITP/001/18TP. We would also like to thank the team at Tory Burch LLC for sharing their insight and ideas regarding existing industry sustainability initiatives, which helped inform some of the references in this work (but are not accountable for any potential errors found in this work).

DOI: http://dx.doi.org/10.5772/intechopen.87062

Author details

Jennifer Xiaopei Wu and Li Li*
Hong Kong Polytechnic University, Hong Kong

*Address all correspondence to: li.lilly@polyu.edu.hk

References

[1] Ellen MacArthur Foundation. A new textiles economy: Redesigning fashion's future. 2017. Available from: https://www.ellenmacarthurfoundation.org/publications

[2] Lehmann M, Tärneberg S, Tochtermann T, Chalmer C, Eder-Hansen J, Seara JF et al. Pulse of the fashion industry report 2018. 2018. Global Fashion Agenda and The Boston Consulting Group. Available from: https://www.globalfashionagenda.com/initiatives/pulse/#

[3] Textile Exchange. 2018 Preferred Fiber and Materials benchmark insights report. 2018a. Textile Exchange. Available from: https://textileexchange.org/downloads/2018-preferred-fiber-materials-benchmark-insights-report/

[4] World Bank. Euromonitor international apparel and footwear 2016 edition (volume sales trends 2005–2015). 2017. World development Indicators—GD

[5] The Business of Fashion and McKinsey and Company. The State of Fashion 2019. 2019. The Business of Fashion and McKinsey and Company. Available from: https://www.businessoffashion.com/articles/intelligence/the-state-of-fashion-2019

[6] The Economist (Producer). Fashion's toxic threads. [Video]. 2018. Available from: https://www.youtube.com/watch?v=Ko_BZhIpI1Q

[7] UK Parliamentary Environmental Audit Committee. Fixing fashion: clothing consumption and sustainability. 2019. UK Parliament House of Commons. Available from: https://publications.parliament.uk/pa/cm201719/cmselect/cmenvaud/1952/report-files/195202.htm

[8] Napper IE, Thompson RC. Release of synthetic microplastic plastic fibres from domestic washing machines: Effects of fabric type and washing conditions. Marine Pollution Bulletin. 2016;**112**(1):39-45. DOI: 10.1016/j.marpolbul.2016.09.025

[9] Global Fashion Agenda. 2020 Commitment. 2018. Available from: https://www.globalfashionagenda.com/commitment/

[10] Waste and Resources Action Programme. Sustainable clothing action plan (SCAP). 2018. Available from: http://www.wrap.org.uk/sustainable-textiles/scap

[11] Kirchain R, Olivetti E, Miller TR, Greene S. Sustainable apparel materials. 2015. Cambridge, MA: Materials Systems Laboratory, Massachusetts Institute of Technology. Available from: https://globalcompostproject.org/wp-content/uploads/2015/10/Sustainable ApparelMaterials.pdf

[12] Schlanger Z. Your Cotton Tote is pretty much the worst replacement for a plastic bag. QUARTZ. 2019. Available from. https://qz.com/1585027/when-it-comes-to-climate-change-cotton-totes-might-be-worse-than-plastic

[13] Sustainable Apparel Coalition. The Higg Index. 2019. Sustainable Apparel Coalition. Available from: https://apparelcoalition.org/the-higg-index/

[14] Waste and Resources Action Programme. Valuing Our Clothes: The Cost of UK Fashion. 2017. Available from: http://www.wrap.org.uk/sustainable-textiles/valuing-our-clothes

[15] Common Objective. Made-By Environmental Benchmark for Fibres. 2018. Common Objective, Available from: https://www.commonobjective.

co/article/made-by-environmental-benchmark-for-fibres

[16] Textile Exchange. Corporate Fiber and Materials Benchmark. 2018b. Textile Exchange. Available from: https://textileexchange.org/pfm-benchmark/

[17] Laitala K, Klepp IG, Henry B. Does use matter? Comparison of environmental impacts of clothing based on fiber type. Sustainability. 2018; **10**(7):6-8. DOI: 10.3390/su10072524. ISSN 2071-1050; CODEN: SUSTDE

[18] Ramot Y, Haim-Zada M, Domb AJ, Nyska A. Biocompatibility and safety of PLA and its copolymers. Advanced Drug Delivery Reviews. 2016;**107**: 153-162. DOI: 10.1016/j.addr.2016. 03.012

[19] Barnes O. Sustainable Yarn. Textiles and Knitwear Team WGSN Materials, WGSN Reports. 2018. WGSN. Available from: https://www-wgsn-com.ezproxy.lb.polyu.edu.hk/content/board_viewer/#/80459/page/1

[20] Dissanayake G, Perera S. New approaches to sustainable fibres. In: Muthu SS, Gardetti M, editors. Sustainable Fibres for Fashion Industry: Volume 2. Springer Singapore: Singapore; 2016. pp. 1-12

[21] The Hong Kong Research Institute of Textiles and Apparel (HKRITA). Inaugural Deployment of HKRITA's Award-Winning Textile Recycling Technologies Helps Close the Loop for the Fashion Industry [Press release]. 2018. Retrieved from: http://www.hkrita.com/marketing/PressRelease/20180903_HKRITA_Media_event_on_Textile)Recycling_Eng.pdf

[22] Unifi Inc. REPREVE. 2019. Available from: https://repreve.com/

[23] Ütebay B, Çelik P, Çay A. Effects of cotton textile waste properties on recycled fibre quality. Journal of Cleaner Production. 2019;**222**:29-35. DOI: 10.1016/j.jclepro.2019.03.033

Chapter 2

Does Mass Customization Enable Sustainability in the Fashion Industry?

D.G.K. Dissanayake

Abstract

Fashion industry evolves today as one of the largest yet among the top of the most polluted industries in the world. Fashion has become cheap and affordable; hence, the consumption has risen to an unsustainable level. Water and energy consumption, hazardous chemical usage, resource depletion, and waste generation are among the key environmental impacts created by the fashion industry. To foster the sustainability in the fashion industry, development of new business models that minimize the environmental damage is urged. This chapter reviews the possibility of a mass customization strategy to become a sustainable business model in the fashion industry. Seven key elements that could possibly enhance sustainability are discussed, and it is concluded that further advancement of technologies and growing consumer desires to purchase sustainable products will make mass customization a viable sustainable business model.

Keywords: fashion industry sustainability, mass customization, personalized fashion, sustainable fashion

1. Introduction to mass customization

In today's fashion market, consumer's desire has been changed from purchasing a mass-produced apparel to a more personalized garment with the right fit and preferred design. This growing desire of product personalization demands apparel manufacturers to move away from mass manufacturing to mass customization [1]. Mass customization in the apparel industry refers to producing a personalized style by adopting individual consumer taste, at the right time and at the right cost. In mass manufacturing, high volumes of identical items are produced, yet in mass customization, unique items are produced for individual customer desires [2, 3] for a relatively large market, yet with efficiency comparable to mass manufacturing [4, 5]. Mass customization is also described as a technology-assisted production process where customers are given the opportunity to modify the traditional mass production process to produce their preferred design and fit [6]. Glimore and Pine define this as a collaborative approach where the manufacturer customizes a product based on customer desires identified through a proper dialog [7]. Davis first brought up the concept of mass customization; emphasizing new technologies will facilitate manufacturing customized products in mass basis [8]. As Nayak et al., argue, apparel purchasing is rapidly moving out of the physical domain into the virtual domain, and technological advancements in the fashion industry such as virtual prototyping, 3-D body scanning, and computer aided design/ manufacturing (CAD/CAM) have already brought mass customization into a reality [2].

Making personalized garments is not entirely a new concept in apparel manufacturing. In pre-industrialized economy, garments were custom-made, catering individual consumer requirements. Measurements were taken from each consumer and the garments were made as 'one-off' pieces with a preferred style and fabric. Consumers used to make their own clothes at home or obtained the service of a tailor to make their garments with their design choice. However, making personalized garments was time-consuming, incurred high unit cost and was not efficient consumption-wise. Furthermore, when getting the service of a tailor, customer had to wait days or weeks to receive the finished garment.

Industrialization made a massive impact to the way garments were made. Production systems were developed to mass produce standard designs with standard sizes at low unit cost. This mass manufacturing strategy facilitated a cost-effective and efficient way of manufacturing garments in a shorter period of time than manufacturing one-off pieces. This led the customers to move away from making personalized garments to the purchase of mass-produced fashion at affordable prices. This eliminated the customer waiting time, as mass-manufactured fashion clothing was ready available to purchase over the counter. However, in the mass-manufacturing process, continuous production run of high volumes made the customization impossible [8], and therefore customers ended up having only few styles in high volumes and different size ranges. Yet, mass-manufactured fashion clothing rapidly captured the consumer market as customized clothing could not compete with the cost and time.

2. Unsustainability of the fashion industry

Mass production and consumption in the fashion industry brought up numerous sustainability issues along the product life cycle. Mass manufacturing enables fashion clothing at affordable prices, which led to increase the consumption and disposal rates [9]. Increasing world population and ever-changing consumer tastes also fuelled unsustainable production, consumption, and throw-away practices [10]. The fashion industry today is characterized by fast-changing fashion cycles, high volumes of production, overconsumption, and frequent disposal habits [11]. To fulfill the ever-increasing demand for fashion products in today's world, natural resources are being consumed in a fast rate than the time required for them to regenerate. Consumption of natural resources such as massive volumes of water, petroleum-based fiber production, use of hazardous chemical, and energy consumption have already made nonreversible impacts in terms of depletion of resources and creating environmental pollution [12]. Offshore production led to increase the energy consumption, transportation emissions, uncontrolled waste generation, and adverse environmental and social impacts, mostly in the least developed countries where manufacturing plants are located [13, 14]. It is estimated that the fashion industry would contribute to a quarter of world carbon budget by 2050, if the current phase of production and consumption continues [15].

Fashion industry today is listed as one of the most polluting industries in the world, therefore experiencing an increasing pressure to integrate sustainability into its supply chain [16, 17]. Industry is already working on adopting sustainability concepts and strategies, yet the integration of new sustainable business models is urged as one promising path to enhance sustainability. In this scenario, mass customization has been identified as one such viable strategy that carries the potential to create both economically and environmentally sustainable business models [18, 19].

3. Enabling sustainable fashion through mass customization

A growing number of apparel manufacturing companies have shown their interest to adopt a mass customization strategy [20, 21]. This is a promising approach in overcoming some of the sustainability challenges inherited in the mass manufactured business model, yet the producers still show their interests only from an economic viewpoint. However, careful analysis of the characteristics of mass customization provides positive insights into reducing excess production, overconsumption, extending product life, and minimizing waste generation. Therefore, it is vital for the manufacturers and retailers to look into this strategy not only in an economic perspective, but also in environmental and social perspectives as well. The following section briefly discusses several possibilities that mass customization may foster sustainability in the fashion industry.

3.1 Improved relationship with the product

Several authors have highlighted the requirement of improving customer attachment to the product, in order to slow down unsustainable consumption and disposal habits [22–24]. Mass customization offers an improved customer relationship to the product and the producer. Each customer is personally treated, which helps to retain a customer long term through the development of trust and relationship between the supplier and the customer. In the mass manufacturing process, the customer is not directly connected with the manufacturer, and their products are made based on general market research and trend information [25]. Nevertheless, in the process of customization, the customer becomes the co-designer, who actively involves in the product development process [26]. This offers the customer a sense of belongingness and attachment to the product and process, even before purchasing the product. This process of involvement could probably replace the joy of customer shopping experience and provides a higher level of satisfaction with the total experience of creating and purchasing personalized fashion clothing. Customer builds up a positive relationship with the product, which ultimately keeps the customer attached to the product for a longer period of time than a mass-produced garment, in which the time between purchasing and disposal has become just a matter of weeks.

The mass customization strategy creates an online platform for the customers to visualize while creating the design and make changes until they are satisfied with the final look. One drawback of online purchasing of clothing in general is highlighted as the inability to confirm the fit before the purchasing decision is made. Digital advancement in the customization process allows overcoming this issue by facilitating the customer to view the design and fit of the garment before making the purchasing decision. 3-D body scanning and virtual prototyping technologies enable online fit sessions even before the garment is physically being made; thus, customer satisfaction is ensured as early as the product development stage. The advancement of technology embedded with the mass customization process is such that the customer can try out few different style changes virtually and observe the fit and style, before making the purchasing decision. This experience indeed brings higher customer satisfaction than visiting a shop and trying out a garment where there is no opportunity to alter based on customer desires. Therefore, the mass customization strategy promotes a better relationship between the customer and the product, starting from the design stage itself. This relationship ultimately keeps the customers happy about their purchasing decision and the product.

3.2 Extended user phase

The time period a garment is used by the consumer depends on various complex factors; yet the degree of consumer attachment to the product is definitely a one key factor, as discussed in Section 3.1. During the customization process, unlike in mass manufacturing, customer attachment to the product is already built up with the involvement of the customer as a co-designer. This sense of ownership and belongingness, together with the uniqueness of the product, make the consumer keeps the garment for a longer period than a mass-manufactured garment with a standard design.

Moreover, mass-manufactured cheap clothing has increasingly become disposable commodities over the last decade. Herein, fashion clothing is meant to be bought, worn once or twice, and quickly disposed of. Availability of clothing in mass scale at affordable prices pushed consumers for unsustainable purchasing and consumption behaviors and invited them to become a part of a throw-away culture. Fast cycle fashion products tend to be manufactured in such a way that the consumers are pushed for short product life cycles, where the quality is compromised for low cost. This led to create high volumes of waste generated in monthly or even weekly basis, pausing environmental threats. Even though it has been emphasized that improving the product quality and increasing selling prices would slow down the damage, growing competition among retailers to increase the sale volumes pushed them for a price competition, where clothing has become cheaper and affordable across consumer segments than ever before. Consumption has become a habit, and clothing is bought for fun, entertainment, and even for single use; thus, long term attachment to the product appears to be impossible to achieve [27].

In contrast, many researchers highlighted the customers' growing desires to pay a premium price for a long-lasting, personalized product [28, 29]. This is mostly associated with the consumers' willingness to express their individuality through what they wear. When a customer pays a premium price for such a personalized product, it is no longer fallen under the category of disposable commodity. Moreover, the product is carefully thought through in the design stage and the price reflects the design and quality, and thus customer satisfaction regarding the product becomes high. This satisfaction motivates the consumer to pay additional price, keep the garment for a longer time than usual, and avoids frequent purchasing of cheap, mass-produced and low-quality clothing. Therefore, through the mass customization strategy, there is a possibility of extending the use phase of the garment life and reducing the environmental burden created by ever-increasing resource extraction for fast phase production.

3.3 Waste minimization

Cheap, industrialized mass production converted fashion into a fast and disposable commodity. The concept of fast fashion is described as a quick response system that caters ever changing, yet uncertain market demands, by offering highly fashionable yet cheap, low-quality clothing with short life cycles [22, 23]. While this drives the profitability of the fashion business, ever-increasing consumption and disposal of clothing poses environmental challenges. In fast fashion phenomena, clothing is made for single use and then thrown away; thus, the use of low-quality materials avoids the possibility of reuse or recycling. Moreover, due to the increasing use of synthetic materials with fiber blends, clothing wastes become neither biodegradable nor recyclable.

Even though frequent purchasing of new clothing keeps the customer emotionally satisfied for some time, the attachment to the product fades away soon due to fit, quality issues, and also the fact that new fashion items are available to purchase almost every week. Moreover, wide availability of similar styles in mass-manufactured system does not bring the consumer the emotional satisfaction of having a unique product. Therefore, the garment is thrown away as new or after single use, generating significant volumes of clothing waste. Global production of clothing has doubled over last 15 years, yet half of the purchased items were thrown away within less than a year [15].

There is a growing desire for customized apparel due to customer dissatisfaction of purchasing mass-produced garments due to fit, quality, or design issues. Mostly the customer has to compromise between fit and style, which results in the garment hanging in the wardrobe without any use or throwing away within a shorter time period than that was interned to keep. Mass customization could dramatically reduce purchasing of apparel that customer is not fully satisfied with. As discussed above in Sections 3.1 and 3.2, personalized styles with right fit and quality would extend the use phase of the product. Customer desire to follow fast fashion trend and throw away culture will be reduced due to customization option. Slowing down the rate of purchase of mass-produced apparel will reduce the demand for fast fashion.

Nevertheless, by actively engaging in the customization process, the customer could engage in a different level of thinking process and also gain a design experience, which may provide better satisfaction than a time-consuming, wasteful, and sometimes an unhappy shopping experience. Once the customers start enjoying the engagement in the customization process, their random purchase of mass-manufactured apparel will be reduced. Reduction in the purchase of unwanted apparel or less fitted apparel means the reduction in disposal and waste generation. In mass production, quantities are based on predicted market demand, and therefore forecasted sales figures are not 100% accurate. This causes overproduction and unsold stocks in the sales store, resulting in waste generation. In mass customization, production is based on actual demand, and therefore, unnecessary production and unsold stocks can be eliminated. This strategy facilitates minimizing wasteful resource consumption and excess production, which in turn reduces the environmental burden.

In mass manufacturing, stock holding cost is high as goods are produced few months in advance and also distribution inventories are added up to the total stock-holding cost. Yet, mass customization offers on-demand production where goods are produced after the order has been placed [26]. Therefore, raw material inventories, post-production, and distribution inventories could be kept to a minimum. Moreover, unsold stock can be eliminated as production is based on actual orders from the end consumer. This helps to reduce high inventory cost and store space.

3.4 Enabling eco-friendly printing technologies

Traditional printing techniques in the mass manufacturing process are time consuming, resource-intensive, polluting, and wasteful [2, 30]. Yet the digital printing technology has the potential to reverse the trend to a more eco-friendly printing solution [30]. With the mass customization strategy, there is a possibility to replace conventional industrial printing with digital printing technologies.

Digital printing facilitates printing individual and unique designs, only according to the order. Digital printing technologies such as inkjet or sublimation printing are less time-consuming, flexible, and offer a cost-effective solution for customized printing. Gupta highlighted the benefits of inkjet printing as personalization and quick response, ability to copy the original design fast, ability to print already made products, environmentally friendly production process, and cost saving due to short fabric lengths and reduced stocks [30].

Digital printing offers the customization in the printing stage where completely new prints can be incorporated for each individual garment. Thus, the customer gets two stages to personalize the product, first in design stage and then in printing. This is a better way to customize the product while minimizing the environmental impact of printing. This option saves significant amounts of energy, water, and chemicals, compared to mass-scale industrial printing such as rotary or screen printing.

3.5 Enabling repair, reuse, or recycling models

As the mass customization process builds up a close relationship with the consumer, implementation of product service systems and product take back systems would be a possibility. There is a growing emphasis on extending the use phase of clothing by offering after sales services such as repairing, where a consumer can bring the product to a service center for repairing or upgrading. When the consumer is no longer interested about the product and wants to dispose, this can be done though a product take back system with some incentive given to the consumer for returning the product, such as discounts over the next purchase. Benefits of product take back systems can be gained by convincing the consumer during the product design stage to use the materials that can be recycled. Returned products can be reused or recycled to minimize the product's end of life environmental burden. With the mass customization strategy, all those sustainable end-of-life product options that are being extensively discussed in general mass production system, and viewed as challenges so far, can come into a reality.

3.6 Enhancing consumer awareness of sustainable fashion

When customer becomes the co-designer of the product, environmental impacts of their product decisions can be easily and effectively communicated during the design stage. This can create a platform for discussion on developing an environmentally responsible product though an effective dialog among the producer and customer. The producer can make the consumer mindful about product sustainability and influence on how to become an environmentally responsible consumer by selecting sustainable production, consumption, and disposal strategies.

Integrating sustainability into the fashion design process has been identified as the most effective way of making a sustainable fashion product [31, 32]. In the mass customization process, this can be facilitated through an integrated sustainability indicator tool within the collaborative design process. Integrating a user friendly simple software tool to indicate the sustainability impact of their choice of material, design, colors, and recycling options can influence the consumer to make sustainable choices in the design stage. This type of an interface would bring benefits to the producer and consumer as the sustainability choices are made through a mutual understanding among both parties. This results in an effective collaborative sustainable design exercise, rather than producer making a sustainable fashion piece and trying to convince the customer to purchase that without understanding the sustainably impacts of their purchasing decision.

3.7 Moving from global to local production

Mass customization is characterized by customer-centric production, competitive cost of manufacturing, and timely delivery of finished goods. In this scenario, to facilitate maximum flexibility in production and short product development cycles, manufacturing base has to be moved away from mass-scale offshore production centers to more local-based manufacturing [33]. The current offshore production system brings not only an environmental cost, but also a social cost. Cheap labor, poor working conditions, low wages, and extended working hours are among the key social impacts of the current system. The absence of environmental regulations allows these developing nations to use hazardous chemicals which are banned in EU countries and to continue water and land pollution by releasing both effluent and solid waste to the environment without proper treatments.

Shifting manufacturing operations back to the countries where majority of the consumption is taking place can reduce the extensive environmental and social damage caused by the global supply chain. Local production has to be adhered to the country-specific environmental regulations and labor laws. This helps the consumer to understand the impact of their consumption behaviors to their environment and the society. Moreover, local production facilitates better control of production and quality, better working conditions, short lead times, and local employment opportunities and reduces the cost and emissions associated with global scale transportation.

4. Discussion

Analysis of the mass customization strategy in environmental and social perspectives highlighted some positive facts towards supporting sustainability efforts of the fashion industry. This strategy facilitates user-maker connection, and therefore builds a close customer relationship with the producer and a deep satisfaction with the product. This helps to extend the garment life and to consume products more responsibly than now in the case, while reducing the need for frequent purchasing, thus slowing down the consumption. Manufacturers move away from product-centric to customer-centric approach, where customer plays an important role in the decision-making in the product development process. The best place to integrate sustainability into any product is the design stage, rather than looking to resolve environmental burdens at the end of pipe line. A collaborative dialog among the customer and the producer during the design development stage can lead to create an environmentally conscious product. Apart from environmental and social benefits, mass customization brings profit to the business in long run through minimizing excess production, reducing inventories and waste generation. Satisfied customers with a particular producer continue their purchasing with the same company; thus loyalty builds up and customer switching between companies will be minimized. Shifting manufacturing base to local plants can dramatically cut down the cost associated with global supply chains and also lead times.

The mass customization strategy is already being adopted by various manufacturing industries such as automotive, computer, electronics, and clothing. Profitability in adopting mass customization is promising as many companies are reported to have double digit sales growth [34]. Similarly, fashion companies have shown their desire on shifting from mass manufacturing to the mass customization strategy. For example, companies such as Levi Strauss, Second Skin Swimwear, Nike, Addidas, and Puma have successfully incorporated mass customization into their businesses [2, 20]. Levi Strauss started offering customized jeans in 1995, and now expanded to a broader range of styles, colors, and fabrics. Nike ID program allows the consumer to customize a pair of shoes using thousands of color combinations and own embroidery designs [2]. Proper Cloth, a US-based apparel company, offers customized men's shirts with the choice of fabric, color, and component design. Famous fashion brands such as JC Penney, Ralph Lauren, and Land's End already offer a variety of customized products with a choice of fabric, color, and design while maintaining order to delivery time of 4–5 weeks [35].

Various consumer studies have reported a positive trend towards purchasing customized apparels. According to the group consumer survey conducted by Taieb and Cheikhrouhou [36], 82% of the teenagers and 92% of the youth and adults have voted for customized products over mass manufactured products. Deloitte consumer review [37] conducted in the UK shows that 41% of the consumers have expressed their interest in purchasing customized clothing, and 19% of the respondents have already bought customized clothing. According to YouGov consumer survey conducted in the US for internet users, 29% of the respondents indicated that they have already purchased customized apparel or footwear [38]. Those studies discovered the growing awareness among consumers regarding customized products and the related benefits. They are also willing to pay a premium price for a customized product. According to the consumer research conducted by Deloitte, 88% of the survey participants are willing to pay an extra price for customized clothing [37]. A total of 67% of the YouGov survey respondents indicated their willingness to pay a premium price for a customized fashion [38]. Most importantly, consumers will benefit through price customization as the price of the product is dependent upon the degree of customization [26]. These consumer researches indicate that price is not a barrier to adopt mass customization into the fashion business.

Mass customization is so far viewed by researchers and practitioners through economic and consumer perspectives, and the literature provides little evidence on environmental or social benefits of the strategy. It is expected that this foundation study would provide a platform for the academia and industry to view mass customization from a sustainability viewpoint. A full life cycle analysis would be useful in quantifying the environmental impact of a customized product against a mass manufactured product, by considering the seven key sustainability factors discovered in this chapter.

5. Conclusion

This chapter offers an insight into the sustainability aspects of the mass customization strategy. Mass customization is becoming a growing trend in the fashion industry due to the changing consumer desires; yet the strategy is already proven to bring profit and sustainability into the business. Digital technologies associated with mass customization facilitated an effective dialog between the producer and the customer while completely cutting down the human travelling or sample transportation during the product development process. Major sustainability benefits of the strategy are identified as improved relationship between the product and the consumer, extended user phase of clothing, waste minimization, enabling eco-friendly printing technologies, enabling repair, reuse, or recycling models, enhancing consumer awareness regarding sustainable fashion, and moving from global to local production. Even though mass customization still offers challenges to the producer in the operational phase, further advancement of technologies and growing consumer desires on personalized products would make this a viable business model that brings positive impacts to the business in economic, environmental, and social perspectives.

Author details

D.G.K. Dissanayake
Department of Textile and Clothing Technology, University of Moratuwa, Sri Lanka

*Address all correspondence to: geethadis@uom.lk

References

[1] Fogliatto FS, Da Silveira GJ, Borenstein D. The mass customisation decade: An updated review of the literature. International Journal of Production Economics. 2012;**138**:14-25

[2] Nayak R, Padhye R, Wang L, Chatterjee K, Gupta S. The role of mass customisation in the apparel industry. International Journal of Fashion Design, Technology and Education. 2015;**8**(2):162-172

[3] Lee SE, Kunz GI, Fiore AM, Campbell J. Acceptance of mass customisation of apparel: Merchandising issues associated with preference for product, process and place. Clothing and Textile Research journal. 2002;**20**:138-146

[4] Piller F. Mass Customisation. 3rd ed. Wiesbaden: Gabler; 2003

[5] Tseng M, Jiao J. Mass Customisation. In: Sakvendy G, editor. Handbook of Industrial Engineering. 3rd ed. New York: Wiley; 2001. pp. 684-709

[6] Fralix MT. From mass production to mass customization. Journal of Textile and Apparel, Technology and Management. 2001;**1**:1-7

[7] Gilmore J, Pine J. The four faces of mass customisation. Harvard Business Review. 1997;**75**(1):91-93

[8] Davis SM. From future perfect: Mass customizing. Planning Review. 1989;**17**(2):16-21

[9] Dissanayake G, Sinha P. An examination of the product development process for fashion remanufacturing. Resource, Conservation & Recycling. 2015, 2015;**10**:94-102

[10] Tojo N, Kogg B, Kiørboe N, Kjær B, Aalto K. Prevention of Textile Waste. 2012. Available from: http://norden.diva-portal.org/smash/get/diva2:701022/FULLTEXT01.pdf

[11] Niinimäki K, Hassi L. Emerging design strategies in sustainable production and consumption of textiles and clothing. Journal of Cleaner Production. 2011;**19**(2011):1876-1883

[12] Allwood J, Laursen SE, de Rodríguez CM, NMP B. Well Dressed? The Present and Future Sustainability of Clothing and Textiles in the United Kingdom. University of Cambridge Institute for Manufacturing; 2006

[13] Dissanayake DGK, Weerasinghe DU, Wijesinghe KAP, Kalpage KMDMP. Developing a compression moulded thermal insulation panel using postindustrial textile waste. Waste Management. 2018;**70**(2018):356-361

[14] Christmann P, Taylor G. Globalization and the environment: Determinants of firm selfregulation in China. Journal of International Business Studies. 2001;**32**(3):439-458

[15] The Guardian. Is fast fashion giving away to the sustainable wardrobe? 2019. Available from: https://www.theguardian.com/business/2018/dec/29/fast-fashion-giving-way-sustainable-wardrobe

[16] Birtwistle G, Moore CM. Fashion clothing—Where does it all end up? International Journal of Retail & Distribution Management. 2007;**35**(3):210-216

[17] Goworek H. Social and environmental sustainability in the clothing industry: A case study of a fair trade retailer. Social Responsibility Journal. 2011;**7**(1):74-86

[18] Pollard D, Chuo S, Lee B. Strategies for mass customisation. Journal of Business & Economics Research. 2008;**6**(7):77-86

[19] Badurdeen F, Iyengar D, Goldsby TJ, Metta H, Gupta S, Jawahir IS. Extending total life-cycle thinking to sustainable supply chain design. International Journal of Product Lifecycle Management. 2009;**4**(1):49-67

[20] Lee S, Chen J. Mass customization methodology for an apparel industry with a future. Journal of Industrial Technology. 1999;**16**:2-8

[21] Loker S, Oh YJ. Technology, customization and time-based performance in the apparel and sewn products industry. Journal of Textile and Apparel, Technology and Management. 2002;**2**:1-13

[22] Gwilt A, Rissanen T. Shaping Sustainable Fashion: Changing the Way We Make and Use Clothes. London, UK: Earthscan; 2011

[23] Fletcher K, Grose L. Fashion & Sustainability. London, UK: Lawrence King; 2012

[24] Niinimäki K. Sustainable Consumer Satisfaction in the Context of Clothing. Product Service System Design for Sustainability. Sheffield, UK: Greenleaf; 2014. pp. 218-236

[25] Moser K, Muller M, Piller FT. Trasforming mass customisation from a marketing instrument to a sustainable business model at Adidas. International Journal of Mass Customisation. 2006;**1**(4):463-471

[26] Piller FT, Müller M. A new marketing approach to mass customisation. International Journal of Computer Integrated Manufacturing. 2004;**17**(7):583-593

[27] Karell E. Planned continuity—Multi-life garments through modular structures & supplemental services. In: Niinimäki K, editor. Sustainable Fashion: New Approaches. 2013. pp. 112-123

[28] Franke N, Keinz P, Steger CJ. Testing the value of customization: When do customers really prefer products tailored to their preferences? Journal of Marketing. 2009;**73**(5):105-121

[29] Grimal L, Guerlain P. Mass customization in apparel industry—Implication of consumer as co-creator. Journal of Economics & Management. 2014;**15**:105-121

[30] Gupta S. Inkjet printing—A revolutionary ecofriendly technique for textile printing. Indian Journal of Fibre and Textile Research. 2001;**26**:151-161

[31] Kerr J, Landry J. Pulse of the Fashion Industry. 2017. Available from: https://globalfashionagenda.com/wp-content/uploads/2017/05/Pulse-of-the-Fashion-Industry_2017.pdf

[32] Boeijen Av, Daalhuizen J, editors. The Delft Design Guide. Delft: The Faculty of Industrial Design Engineering, Delft University of Technology; 2010

[33] Ulrich PV, Anderson-Connell LJ, Wu W. Consumer co-design of apparel for mass customisation. Journal of Fashion Marketing and Management. 2003;**7**:398-412

[34] Piller FT. Observations on the present and future of mass customization. International Journal of Flexible Manufacturing Systems. 2007;**19**:630-636

[35] Yeung HT, Choi TM, Chiu CH. Innovative mass customization in the fashion industry. In: Cheng T, Choi TM, editors. Innovative Quick Response

Programs in Logistics and Supply Chain Management. International Handbooks on Information Systems. Berlin, Heidelberg: Springer; 2010

[36] Taieb AH, Cheikhrouhou M. Study of mass customisation approach in apparel and textile field. International Journal of Engineering Sciences & Research Technology. 2016;5(1):223-234

[37] The Deloitte Consumer Review. Made-to-Order: The Rise of Mass Personalisation. 2019. Available from: https://www2.deloitte.com/content/dam/Deloitte/ch/Documents/consumer-business/ch-en-consumer-business-made-to-order-consumer-review.pdf

[38] YouGov. Made to order: An analysis of US consumer perception towards personalization. 2018. Available from: https://today.yougov.com/topics/resources/articles-reports/2018/06/07/made-order-analysis-us-consumer-perception-towards

Chapter 3

Exploration of Bamboo Fabrics with Natural Dyes for Sustainability

Kavita Chaudhary

Abstract

The history of weaving and textile innovation follows the history of technological innovations and the evolution of socio-cultural issues in a stepwise manner. Currently, textiles and patterns are tools that characterize garments and define their unique aesthetics, becoming identity-making elements not only for a single product or a collection, but also for brands. By balancing a delicate equilibrium between brand heritage and zeitgeist, the Indian textile industry bases its success on the capability to reorganize creative processes to generate innovation, also thanks to new relationships among textile designers and their consumers. It is creating something new is the need of hour for the apparel industry and they are aware of it. By utilizing these new techniques for bamboo fabric, we can introduce some designs which will give unique look and increase the marketability of the product. Bamboo viscose, also known as regenerated bamboo, is a regenerated cellulosic fiber. In the last few years in the world market, more and more products from bamboo fibers have been appearing. It is marketed as having exceptional properties such as superior comfort and hand, as well as antimicrobial properties. Yarns of bamboo fibers provide the desirable properties of strength, high moisture absorbency, antimicrobial and antifungal, sheen and smooth close to silk, cashmere and glass fiber high breathability and thermo-regulating properties in textiles and made ups. The articles are designed and developed by natural dyeing and printing with hand block.

Keywords: bamboo fabric, natural dyeing, natural printing, hand block print

1. Introduction

Bamboo viscose, also known as regenerated bamboo is a regenerated cellulosic fiber. In the last few years in the world market, many products from bamboo fibers have been appearing.

It is marketed as having exceptional properties such as superior comfort and hand yarns of bamboo fibers provide the desirable properties of strength, high moisture absorbency, antimicrobial and antifungal, sheen and smooth close to silk, cashmere, glass fiber high breathability and thermo-regulating properties in textiles and made ups [1]. Shaw [2] currently, regenerated bamboo fibers are used in apparel including undergarments, sportswear, t-shirts and socks, this fiber will not cause skin allergies and its application in sanitary materials such as baby diaper, absorbent pads and sanitary towels. Bamboo fiber has found its way into the fashion world. In their collection, designers often replace more expensive materials like cashmere, silk and glass fiber by their bamboo equivalent [3, 4].

Bamboo has the fastest growth rate among the various types of renewable natural fibers, bamboo fabrics require less dyestuff than cotton fabrics in order to be dyed to the level desired, as they absorb the dyestuff better and faster and show the colour better [5]. The most important aspects of clothing is comfort properties like thermal resistance, air permeability, water vapor and liquid water permeability are critical for the thermal comfort of a clothed body. Comfort plays a vital role in the selection of apparel [12]. Bamboo fiber is more efficient than the cotton fiber as it needs less dying color and raw material to dye the fabric as per the required need. The natural dyeing solution was obtained by from rind of Vempadam Bark (*Ventilago madraspatna*) used for dyeing bamboo fabric. Fastness properties of the dyed sample were studied [6].

Zakrzewski [7] reported that cross-section of the bamboo fiber is covered with micro-holes giving the fabric better moisture absorption and ventilation. According to authoritative testing figures, apparels made from bamboo fibers are 1–2 degrees lower than normal apparels in hot summer. Apparel made from bamboo fiber is crowned as *air conditioning dress* [8].

Bamboo fibers are claimed to have some distinctive properties such as high moisture absorption, better drape and deep colour effect [9, 10]. Bamboo fiber are claimed to have unique properties such as inherent antimicrobial, UV-shielding properties without aid from petrol-derivative chemicals. These unique properties and the sustainable nature of bamboo fiber have started to attract consumers in the textile market, particularly due to the high price [11].

2. Objectives

1. Comparative study of natural dyed, printed bamboo and cotton fabrics.

2. Developed garments of the natural dyed and printed bamboo fabric according to "Fashion Trend."

3. Methodology

3.1 Material

Bamboo yarns were used count 1/64 Nm both side warp and weft direction.

3.2 Method

Bamboo fabric was developed on the handloom in 100 reed count, 2/dent of yarn with plain weave used bamboo yarns both side warp and weft. After manufacturing following process on bamboo fabric.

3.2.1 Natural dyeing and printing processes

3.2.1.1 Scouring

M:L—1:50
Time—60 min at 60–90°C temperature
Casting soda—5% (by weight of the fabric)
Common salt—2% by weight of the fabric (**Figures 1–3**).

Figure 1.
Scouring.

Figure 2.
Washing by fresh water.

Figure 3.
Drying in open air.

3.2.1.2 Mordanting

Dyeing of pre-mordant of bamboo fabric, pre-mordanting treatment was conducted using mordant. Bamboo fabric samples were pre-mordanted of bamboo was carried out with mordant at concentrations on the weight of fabric:

- *Brightening mordant*:

Alum (aluminum potassium sulfate): 25% by weight of fabric

Tin (stannous chloride): 3% by weight of fabric

Chrome (potassium dichromate): 1% by weight of fabric

- *Dulling mordant*:

Copper (copper sulfate): 3% by weight of fabric

Iron (ferrous sulfate): 5% by weight of fabric

At 60–90°C for 30–60 min with material to liquor ratio as 1:30. After mordanting the samples were rinsed in cold water to remove the excess of mordant and used for dyeing (**Figures 4** and 5).

3.2.1.3 Preparation of dyeing processes

Dye solution was boiling for ½ h at 40°C and filter of the dye bath after that dyeing of the pre-mordanted samples was performed for 1 h at 30–40°C in an open bath beaker dyeing machine at 20:1 liquor to material ratio. The dyed samples were rinsed in cold water and dried in open air (**Figures 6–9**).

3.2.2 Preparation of printing processes

3.2.2.1 Mordanting of bamboo fabric

Mordanting of bamboo was carried out with mordant at concentrations on the weight of fabric alum (aluminum potassium sulfate): 25% at 60–90°C for 30–60 min with material to liquor ratio as 1:30. The mordanted fabric was squeezed and dried in open width form at 100°C for 5 min and then used for printing.

Figure 4.
Mordanting.

DOI: http://dx.doi.org/10.5772/intechopen.88688

Figure 5.
Washing by fresh water.

Figure 6.
Dye solution boiling.

Figure 7.
Filter of the dye solution.

Figure 8.
Fabric dying in dye bath.

Figure 9.
Dyed and stitched garment.

3.2.2.2 Preparation of dye paste

All the dyes used were first converted to powder form. For natural dyes were first dried in an oven at controlled environment at 27 ± 2°C and RH 65 ± 2%, with at least 24 h at "equilibrium regain" and later grinded in mixer grinder. The powder so obtained was filtered through the 60 mesh nylon fabric. The fine powder of natural dye was used for printing (**Figures 10** and **11**).

3.2.2.3 Preparation of guar gum paste (print medium)

Guar gum being nonionic is best suited for textile printing was prepared using 5% of guar gum powder. The guar gum powder was sprinkled slowly in water with

Figure 10.
Mesh of the dye in water.

Figure 11.
Filter of the dye paste.

continuous stirring in order to prevent lump formation. The paste was stirred continuously at 90°C for 1 h. For preparation of 1% of print paste, 1 g of dye powder was first pasted with small quantity of water followed by addition of 99 g of guar gum stock paste. The mordanted bamboo fabrics were then printed with two strokes of squeeze, steamed at 100°C for 10 min. The samples were washed with water and then dried in air (**Figures 12–16**).

3.2.2.4 Block printing: preparations of blocks include designing technique and motifs

See **Figure 17**.

3.2.3 Developed garments of the natural dyed and print bamboo fabric according to fashion trend

See **Figures 18–20**.

3.2.4 Comparative study of the bamboo fabric and cotton fabric

The results of the study are shown in **Table 1**.

Figure 12.
Mesh of the guar gum in water.

Figure 13.
Filter of the guar gum paste.

Figure 14.
Guar gum and dye solution.

Figure 15.
Preparation dye trey.

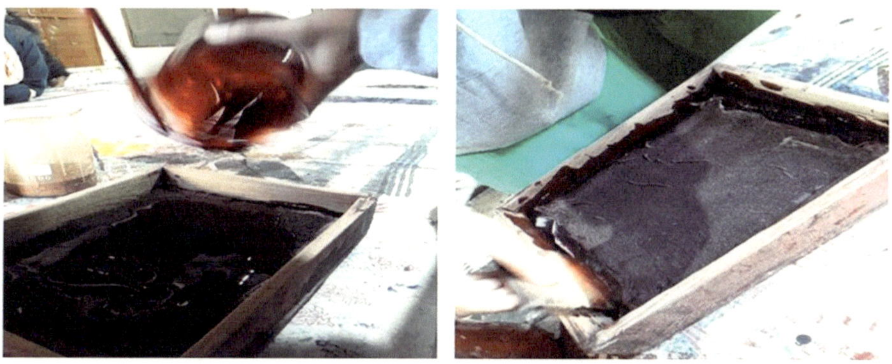

Figure 16.
Preparation of print trey with mixed dye and guar gum paste.

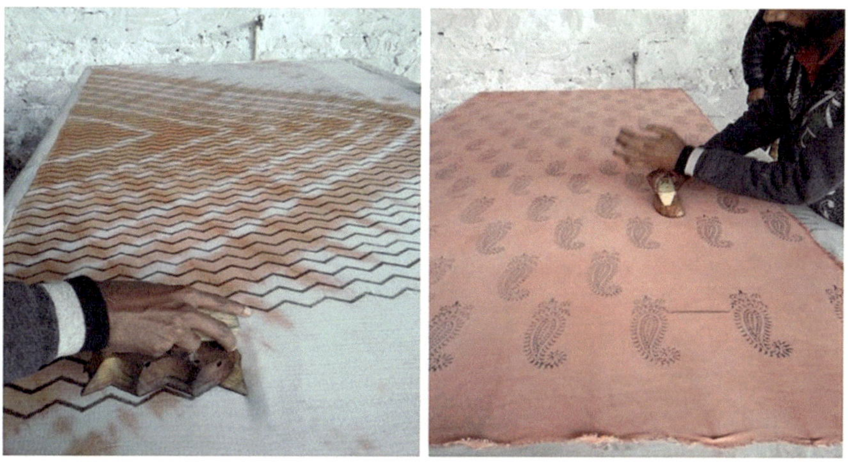

Figure 17.
Printing by wooden blocks.

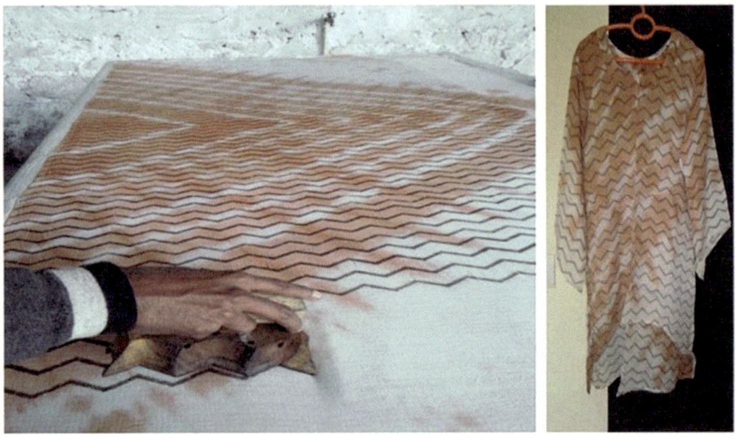

Figure 18.
Onion skin print on dyed fabric with pomegranate peel.

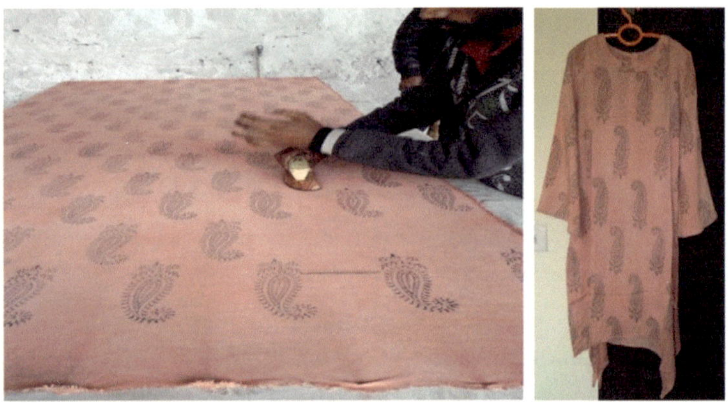

Figure 19.
Onion skin print on dyed fabric with sandal red.

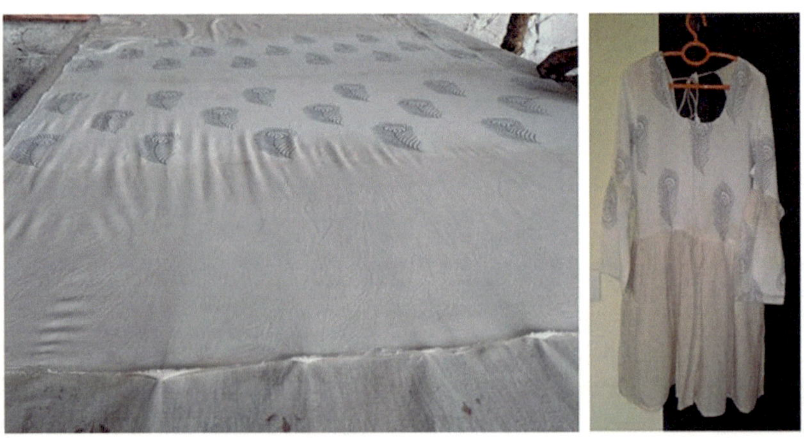

Figure 20.
Indigo blue print on dyed fabric with purging nut.

S. N.	Species (Family Name)	Common Name	PH Value	Colour Obtained	Colour Application	
					Cotton	Bamboo
1.	Caealpinia Sappan	Sappan Wood	5-7	Jaipur Pink	Good	Very Good
2.	Punica granatum-(Lythraceae)	Pomegranate Peel	3-5	Mallow Gold	Fair/Good	Good
3.	Rheum Emido	Indian Ruharb	3-5	Apsara Yellow	Fair	Good
4.	Nyctanthes tristis (Oleaceae)	Night Jasmine	3	Prime Rose Yellow	Good	Very Good
5.	Indigoferatinctoria (Fabaceae)	Indigo	11-12	Indigo Blue	Good	Very Good
6.	Quercus Infectoria	Oak Tree Fruit	5	Gallnut	Good	Very Good
7.	Tagetes Erecta	Mari Gold	5	Sun Yellow	Fair	Good
8	Bixa Orellana	Annatto	9	Candy Orange	Very Good	Excellent
9.	Terminalia che	Myrobalan	5	Cedar Yellow	Good	Very Good
10.	Laccifer laca (keer)	Sheel Lac	5	Wine Red	Good	Very Good
11.	Rubia Cardifolia	Heart Leaved Madder Roots	5-7	Turkey Red	Good	Very Good
12.	Pterocarpus Sandal	Sandal Red	5	Barn Red	Fair/Good	Good
13.	Butea Monosperma	Purging Nut	5-6	Cuttack Silver	Fair/Good	Good
14.	Allium Cepa	Onion Skin	6-8	Onion Peel	Very Good	Excellent

Table 1.
Colour obtained with PH value of the natural dyes by cotton and bamboo fabrics.

4. Conclusion

The experimental research design was used keeping in view the objectives of the study. Pit handloom used by the weaver was employed for bamboo yarns were used to manufacture fabric; handloom bamboo fabric was selected for the study. Fourteen natural dyes used for dyeing and printing with alum mordent. A property such as washing fastness was determined by standard methods. Washing fastness was exhibited fair to excellent in **Table 1** [13]. It can be concluded that creating something new is the need of hour for the apparel industry and they are aware of it. Handloom bamboo fabric was used for apparel in this study considering its dye exhaustion properties. The study is successful innovative bamboo apparel products with natural dyes.

Author details

Kavita Chaudhary
Banasthali Vidyapith, Rajasthan, India

*Address all correspondence to: kavita.bid@banasthali.in

References

[1] Lipp-Symonowicz B, Sztajnowski S, Wojciechowska D. New commercial fibres called 'Bamboo Fibres'—Their structure and properties. Fibres & Textiles in Eastern Europe. 2011;**19**(1):18-23

[2] Shaw M. Bamboo Fibers: A Review [Internet]. 2013. Available from: www.fibre2fashion.com [Accessed: 24 July 2013]

[3] Prakash C, Ramakrishnan G, Koushik CV. Effect of blend ratio on the quality characteristics of bamboo/cotton blended ring spun yarn. Fibres & Textiles in Eastern Europe. 2011;**19**(6):38-40

[4] Saravanan K, Prakash C. Bamboo fibres & their application in textiles. The Indian Textile Journal. 2007;**117**:33-36

[5] Sekerden F. Effect of fabric weave and weft types on the characteristics of bamboo/cotton woven fabrics. Fibres & Textiles in Eastern Europe. 2011;**19**(6):47-52

[6] Manimekalai G. Antimicrobial properties of bamboo of fabric finished with Vempadam bark (*Ventilago madraspantana*) extract. International Journal of Scientific and Engineering Research. 2018;**9**(12):2018

[7] Zakrzewski W. Nowe trendy: A możetak ... bambus? [Internet]. 2011. Available from: http://www.hike.pl/sklep/Nowe-trendy-bambus-sp-25.html

[8] Rodgers M. Bamboo Clothing [Internet]. 2010. Available from: www.fibre2fashion.com [Accessed: 30 April 2010]

[9] Bamboosa. 2009. Available from: https://www.bamboosa.com/boombosa.php?PID=13 [Accessed: 14 January 2009]

[10] Bambrotex. 2008. Available from: http://www.bambrotex.com/second/anti_UV.htm [Accessed: 17 December 2008]

[11] Waite M. Sustainable textiles: The role of bamboo and a comparison of bamboo textile properties. Journal of Textile and Apparel, Technology and Management. 2009;**6**(2):1-21

[12] Chidambaram P, Govindan R. Influence of blend ratio on thermal properties of bamboo/cotton blended woven fabrics. Silpakorn University Science and Technology Journal. 2012;**6**(2):49-55

[13] Saxena S, Raja A. Natural Dyes: Sources, Chemistry, Application and Sustainability Issues. In: Murthu S edition. Roadmap to Sustainable Textiles and Clothing. Textile Science and Clothing Technology. 2014:37-80

Section 2

Trends in Fashion Industry: A Technological Approach

Chapter 4

The Counterfeit Market and the Luxury Goods

Amélia Maria Pinto da Cunha Brandão and Mahesh Gadekar

Abstract

Product counterfeiting is a form of consumer fraud: a product is sold, purporting to be something that it is not. Counterfeit is illegally made products that resemble the genuine goods but are typically of lower quality in terms of performance, reliability or durability. There is a need to learn as much as possible about the counterfeit market and the luxury good's market, as well as learning about the customers of both of these markets. A large number of counterfeits are purchased online as the customer relies only in the information provided by the seller that is easy to manipulate. The Internet makes it possible for counterfeiters to sell fake goods without prior consumer inspection. Thus, if the presentation of the product online is elaborated and advanced, it would influence the user's perceptions of the authenticity of products in a positive way, which leads to an increase in the user's willingness to buy online. The chapter focuses on the counterfeit market, its influence on luxury consumption and consumers' drivers for the counterfeit. The chapter also discusses innovative ways how authenticity of luxury goods can be verified.

Keywords: counterfeit, counterfeit market, luxury goods, fashion goods, anti-counterfeit technology

1. Introduction

The luxury good's market has grown from a value of US$ 20 billion in 1985 [1] to its current value estimated by Bain and Company (2012) of over US$ 260 billion, and the expectation regarding the growth of this market is of about 2 up to 4% annually for the next 2 years [1]. For instance, the high-end luxury good's market came up to US$ 236 billion in 2009 [2]. While the market for luxury goods has grown steadily through the last 20 years representing a significant percentage of trades worldwide, studies have expressed concern about the existence of counterfeit, and generation of loss from counterfeit products amounts to US$ 12 billion per year [3].

Lai and Zaichkowsky [4] define counterfeits as illegally made products that resemble the genuine goods but are typically of lower quality in terms of performance, reliability or durability [5]. Counterfeit goods decrease companies' profits because what these manufacturers' sell is what the brands will not sell, which means that there is a direct steal of revenues from brands. This industry suffers with counterfeits resulting in seizures of large number of counterfeit goods. Thus, the increase in demand of luxury goods is the driver of the demand for counterfeits [6].

The need for individualism and uniqueness is one of important characteristics of counterfeit consumers. Studies have stated that people feel the need for individualism and uniqueness [2, 7], and this necessity can be expressed by means of personal brand preference and buying decisions [8]. Since the idea that money acquires status, which can be seen through the goods people own, people started to acquire products not only for 'what they can do, but also for what they mean' [9] (p. 118). In addition, there is also meaning attached to goods, especially to luxury goods as with the ownership comes a sense of exclusivity. Thus, buying counterfeit goods translates into reality as a loss of exclusivity for the customers of genuine luxury brands [10].

Therefore, this chapter focuses on the counterfeit market, its influence on luxury consumption and consumers' drivers for the counterfeit. Furthermore, the chapter discusses the innovative ways adopted by a Portuguese firm to control the growth of counterfeit market. In particular, start-ups are advised to plan, develop and implement a strategy for launching an anti-counterfeit technology product. However, reaching the customers with anti-counterfeit technology products is not easy and requires focus across the industry. Knowledge of the study helps to understand to control the growth of counterfeit market and loss of sales and image to the fashion firms.

2. Literature review

2.1 Counterfeits and counterfeit market

'Product counterfeiting is a form of consumer fraud: a product is sold, purporting to be something that it is not' [11] (p. 173). However there is also another concept that needs to be brought to light in order to provide further understanding on the subject, which is deception. According to [12], deception is a form of manipulating information with the purpose of influencing the behaviour of a customer of potential customer, by misrepresenting a product. [13]. Ekman [14] stated that there are two major deception strategies, which are concealment, where the deceiver hides relevant information without saying anything false, and falsifying, where the deceiver adds untrue information as if it were true in order to manipulate the customer [13]. In both cases, deception is intentional.

While defining the counterfeit market, [15] believes there are two types of markets for counterfeits, deceptive markets, where the customers are imperfectly informed or are not familiarised with the product, which makes them unable to distinguish between genuine and fake products and leads to the purchase of fake goods by accident [7], and nondeceptive, where the customer can easily differentiate fake from authentic, which cannibalises the purchase of genuine products; this sort of market applies more to the luxury goods industry, since consumers are fully aware they are purchasing a counterfeit when it comes to the date of acquisition; here buyers are accomplices rather than victims [13].

OECD [16] came to the estimate that 5–9% of the global trade volume is counterfeit production, which translates to a basis of a yearly volume of US$ 9 billion; this means that the counterfeit industry accounts for a loss between US$ 450 and US$ 810 million [7]. World Custom Organisation [17] computed that the global market for counterfeits is to exceed US$ 600 billion, accounting for approximately 7% of world trade [3]. The estimate from [18] states that the loss from counterfeiting luxury goods or brands amounts to US$ 12 billion per year [3], which means that sales of counterfeit items have caused billions of dollars in lost sales for legitimate luxury good producers [19]. A different source found was the International Chamber of Commerce, which cites a similar estimate where counterfeits account for 5% all the way up to 7% of worldwide trade, which are worth US$ 600 billion yearly [11].

The counterfeiting industry operates in a 'grey market'; this means that solid numbers are hard to estimate and to come by, hence all of the different yet similar numbers stated above; there is a gap in the academia when it comes to recent and updated statistics for the counterfeit market [11].

All of the estimates mentioned above, namely, that counterfeits account for 5–9% of world trades, have remained static for a few years now, and the assessed percentage does not seem to be based upon empirical findings [11]. Keeping this problem in mind, back in 2016 [16], the Organisation for Economic Cooperation and Development reassessed the issue and projected a value that seems more reasonable and accurate for the volume of counterfeits traded globally, which was based on solid evidence; this value came down to 2% [20], which is significantly lower than the previously estimated value. Another one of the most recent statistics belongs to Interpol (2009), where the value released confirms that counterfeits account for 2% of worldwide trades [10], which goes along the lines of the value released by the [16].

Havocscope [21] confirms that the current value of the counterfeit good's market is about US$ 12 billion. Information extracted from [21] provides us with knowledge regarding the amount of US$ lost by sector per year. As can be seen in Table 1.1, in the Appendix, show companies and clothing manufacturers lose up to US$ 12 billion yearly due to illegal counterfeit activity, while the cosmetic sector loses up to US$ 3 billion per year [21].

With the presented information, it can be seen that there is a real loss in the luxury goods market due to the emergence of this 'grey market' of counterfeits. What this means is that luxury brands face a serious problem: counterfeiting. The luxury good's market has grown from a value of US$ 20 billion in 1985 [1] to its current value estimated by Bain and Company (2012) of over US$ 260 billion, and the expectation regarding the growth of this market is of about 2 up to 4% annually for the next 2 years [1]. Although the technical term 'luxury good' is independent of the goods' quality, customers consider these kinds of products to be of the highest quality available in the market, since that is how these types of brands position themselves; they make the extra effort to associate elevated prices with high-quality standards [22].

The global luxury goods industry, which includes drinks, fashion, cosmetics, fragrances, watches, jewellery, luggage and handbags, has been on an upward climb for many years [23], the luxury goods market is significant, not only in terms of its market value, but also in terms of its growth rate [1, 24]. Luxury goods manufacturers meet consumer demand by focusing on brand, aesthetics, quality materials, superior craftsmanship and pricing to transform everyday objects into status symbols. The industry rises and falls with the gross domestic product (GDP), seeing demand climb in times of economic stability and plummeting in unfavourable economic climates [23]. The United States has long been the largest regional market for luxury goods and is estimated to continue to be the leading personal luxury goods market in 2013, with a value of 62.5 billion euros [23].

2.2 Acquiring counterfeit Good's online

One third of counterfeit sales of luxury goods are made online [2, 3]. Also, and according to [25], consumers of counterfeit goods acquire them mostly through the Internet, especially when they are well aware they are buying a fake good, since it is easier for them to analyse all the information provided, as well as compare the fake products with the genuine articles.

According to [26], a massive 87% of counterfeits are acquired online; the sources this website uses are based on empirical evidence, such as news and feedback from online platforms (eBay, AliExpress, Amazon and Facebook). eBay is currently dealing with the highest percentage of counterfeit complaints; however, and according [26], it is the only one of these platforms doing something to prevent and protect their customers against the counterfeit problem, unlike Amazon, who warns customers to buy at 'their own risk'. The Internet makes it possible for counterfeiters to sell fake goods without prior consumer inspection [13]. [27] estimates that US$ 135 billion were sold online in counterfeit alone, back in 2008 [13].

When buying online, the customer relies only in the information provided by the seller—text, pictures or videos—which is easy to manipulate [13]. A study elaborated by [13] showed that if the presentation of the product online is elaborated and advanced, it would influence the user's perceptions of the authenticity of products in a positive way, which leads to an increase in the user's willingness to buy online.

The Internet has made the practice of selling counterfeit goods much easier, since it decreases costs and increases information asymmetry between parties, making it more difficult to prosecute online deceivers [12, 13].

2.3 Drivers of the counterfeit market

There is a need to analyse what guides this parallel market, what is the appeal behind living of an illegal activity. According to the literature, the two main forces driving this market are, firstly, the high margins obtained by counterfeit manufacturers and, secondly, the demand for name-brand goods and lower prices [3].

World Custom Organisation [17] states that the key driver for this market is the exponential growth of China and the intensification of their exports [11]. Naturally not all counterfeited products come from China; however, an astonishing 2/3 of the total counterfeit shipments detected by authorities actually departed from China; this percentage accounts for about 241 million pieces seized globally [11]. US Customs and Border Protection [28] states that top commodities preferred by counterfeiters include footwear, apparel, watches, pharmaceuticals and electronics [13], which confirms that there is a real issue to be studied.

For consumers, luxury has to be seen by everyone; customers acquire an expensive item or a counterfeit item for their own self-esteem and for others to see as well, in order to attain or increase their social status [13]. The interest in purchasing a high-end product decreases if potential customers consider the set of products to be held by the masses [29], because the decision to buy branded products is based upon social benefits, such as the prestige of holding a good of this nature [29–32]; holding a luxury good represents a symbol of success for the customer [32], and any fashion brand needs to maintain its exclusivity in order to attract customers, since if a brand wants to price the products at designer prices, then it means that it cannot be held by too many consumers [29].

Authors have presented a theory concerning two types of social groups, the elites and the masses; the elites seek to distinguish themselves from the masses, hence their high-end purchases, while the masses have a need to emulate the choices of the elites [29]. The primary value of luxury brands is psychological and that their consumption is dependent on a distinctive mix of social and individual cues [1]. Han et al. [33] states that people do not desire counterfeit goods; they desire the real product; however, sometimes customers whom aspire to own luxury products are unable to afford to pay for the genuine product and can use counterfeits as a poor substitute [19], which means that consumers' desire for counterfeit luxury goods always derives from their preference for genuine luxury brands, and counterfeits provide the user with social status at a low price, but also at a low quality [10], which, besides hindering innovation, employment and trade [34], does not really hurt the final customer in any way, hence their continuous purchases.

In 1993, [15] conducted a study that led to the conclusion that people attach more importance to the image that a brand conveys than to the acute lack of quality and functionality of a copy [7]. As expected, people pay more attention to what holding a certain luxury brand really means to others, more so than to the actual quality of the product; they pay more attention to the external rather than the internal attributes of the product; specifically, this factor leads Asian consumers to have a stronger appetite for luxury goods [19]. Consumers buy the fake goods due to the brand status they get from it, counterfeit status goods often offer similar qualities as the original merchandise and the consumer of these goods is not at any risk of bodily harm; however, consumers of these goods subject themselves to social risk, since these goods are of high symbolic value and social visibility [35]. Bloch et al. [15] states that the true driver for the demand for counterfeits is the demand for luxury goods; as stated above, consumers of fake luxury goods are willing accomplices rather than victims of deception [10], in which counterfeit luxury goods stimulate consumer demand for brand-name items, to some extent [19].

Some counterfeit buyers believe they are receiving comparable goods at lower prices; they see nothing wrong with the action and perceive designer prices as unfair [36], instead of thinking about what really goes with designing and assembling a luxury product [10]. Prestige, brand image and fashion are important to purchasers of premium brands; however, buyers of counterfeits value the same, but try to gain the image benefits at a bargain price [37]. Marketers keep focusing on the quality/price ratio: consumers often find they can get sufficient quality for the price they are paying [10]. The transaction of a counterfeit product may seem advantageous to the seller and to the consumer at the time, but it harms the brand being duplicated, since there is a direct steal of revenues.

According to a study performed by [38], the consumer values ethical standards, and when they feel that a company is not living up to those standards, they will sacrifice themselves to punish the company, which could translate into going out and acquiring counterfeits. The opposite can also happen; customers may sacrifice themselves to defend a company with high moral standards [10, 38]. The study concluded that if the price of the original good is much higher than the price of the counterfeit good, the quality is sufficient, and the customer feels no need to support the manufacturer of the original product; then it is more likely that the customer will acquire the counterfeit, since price remains one of the most important indicators in the market of a product's value [39]. which means that consumers that knowingly acquire a counterfeit good are willing to trade up quality and performance for the fake image purchased at a discount price [37].

According to the literature, there are three main reasons that lead people to acquire counterfeits. Firstly, the price of the counterfeit is much lower than the original; secondly, the quality of the fake is comparable to that of the original; and lastly, customers do not see any reason to support the company that manufactures the original good, which relates to the social responsibility of the company [10]. Poddar et al. [10] conducted a study which provided us with two relevant conclusions for this report, namely, if price differentials are higher and quality differentials are lower, then purchase intentions increase, and, second and extremely relevant in the sense that it confirms what has been said previously, customers are more prone to acquiring counterfeits if they feel that other people will not be able to tell the difference [10].

As seen before, counterfeiters are dealing mainly online, and they provide the customer with professional-looking websites, low prices, faster delivery times and a wider range of delivery, and this poses a real threat to luxury houses [35]. Cordell et al. [40] understand that potential deception occurs in commercial transactions due to information asymmetry and opportunistic behaviour [13], and this is easier to occur online, where everything can be manipulated, in terms of information provided by the seller, as well as visual aids. Research by [41] confirms that advanced presentation online affects buyers in a positive way; it leads to the decrease of apparent risk, creates a better mood for customers and provides a more entertaining experience for them [13].

One of the more relevant factors for consumers that purchase counterfeits online is price; usually these consumers use the Internet as a tool for gathering information and even as a means to compare the fake with the original product [8, 35]. Another factor that is valued by customers is safety; when counterfeiters produce goods, they have no regard for the customer's personal safety [37]; however, this does not impact the luxury-counterfeited goods, since most of them are products to wear, instead of medical drugs, for example, where the danger of buying altered goods is more clear to the end-customer [11].

Counterfeit items have some symbolic value for the customers who buy them with the desire of enhancing their social status [1, 19]. According to [18], the desire of consumers for counterfeit luxury goods depends on the extent to which such brands fulfil the social motivation guiding the luxury brand preferences of those consumers [19]. Thus, consumers with strong face consciousness preferred counterfeit luxury goods that can be used in public to those that can be used in private situations, which supports findings by other authors stating that situational effects are important in considered analysis of nondeceptive counterfeit luxury products [42].

2.4 Innovative way to fight off counterfeits

MobilityNow is a start-up company that has its sole focus on mobile solutions. While there are more than 1.2 billion users who already access the Internet through mobile devices and almost half of them use their mobile in the electronic shopping process, MobilityNow is playing a huge role by adapting the technologies. Mobile device is of growing importance nowadays, since more than 15% of all the Internet traffic comes from mobile devices, which means there is a need for companies to adapt their technologies, and that is where MobilityNow plays a huge role.

The MobilityNow was founded in 2012 under design to develop solutions for the mobile world to help people and businesses to create and manage, with security, a strong digital presence. The company believes the online world has been made to be permanently connected to the offline world through mobile platforms and advocates building an ecosystem where people, machines, sites and cities are securely linked and identified.

Although the Internet assists in counterfeiting spread faster and easier, there are several software that can prevent counterfeiting. BULLA is a simple and innovative solution to fight off counterfeits. It consists of integrating a secure near-field communication (NFC) chip in the target product as well as associating this chip with the company authentication server, BULLA-id. It is also described as a short-range wireless communication between electronic devices. The system through android mobile app and product management platform can immediately verify the authentication of the brand. The NFC can be embedded into the products and contain electronically stored information that can be read by mobile devices. One of the benefits that the NFC provides is that they are quicker to read; hence, it is less time consuming and less laborious. It can be read by a smart phone through mobile apps. NFC is the powerful tool that makes use of the latest authentication technologies. It results in providing producers and consumers access to a new world of possibilities that bring them closer and solidify their bond of trust. Thus, BULLA can be integrated into several categories of products.

In fashion industry, BULLA can be integrated to reduce the risk of counterfeiting and protecting the brand—its processes, products and innovations—from fakes and simultaneously reduces companies' losses. For example, brands by placing the BULLA seal on the bottle will help in identifying counterfeit. The users of BULLA are benefitted in three ways—first, BULLA provides the end-user with authenticity validation. Because it works as a direct way for the customer to interact with the brand, it provides an open line of communication with the brand. Lastly, it is easy and simple for the consumer to use and has a user-friendly interface. For example, a user can scan the BULLA tag placed on the brand through downloaded mobile application.

2.5 Strategic phase of reaching consumers

Considering the presence of competitors for anti-counterfeiting technology solutions products, MobilityNow identified the strength and weaknesses of BULLA to provide value that consumers can relate with their problems of counterfeit products. BULLA has two main strong points worth mentioning, such as its simplicity of usage and it being very easy for the final user to understand and utilise; and, the connections MobilityNow have within the wine sector may come to be a strength towards developing partnerships. The company also identified the role of the content marketing.

Content marketing is a special kind of marketing that has recently emerged due to the evolution of technology and the impact this evolution has had over social media. It is a technique that involves the creation and sharing of media and publishing content in order to acquire and retain customers [43]. Content marketing can be thought of as a way of providing relevant information to customers or potential customers, and the set of information has the aim and the power of maintaining or even changing behaviours [44]. A video was prepared to engage with the consumers and help new consumers to stay connected on the website focusing on the intended benefits of BULLA. The BULLA website was clutter free and concise. The website is easy to navigate and gets in touch with us easily in case potential customers are interested, and most important of all, it is simple to understand what BULLA is all about, how it can be used and where it can be applied.

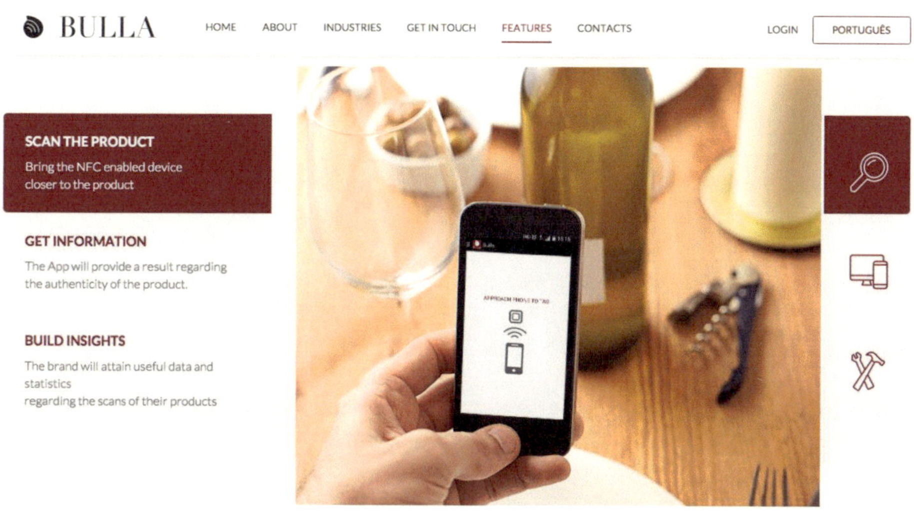

3. Conclusion

When it comes to the luxury goods market, only a fraction of the market acquires these legitimate products, the ones that can afford these goods. The other ones, those who cannot afford it, go for the option of buying a fake, since some of them have the need for these sorts of items but not the availability to acquire them. Most of them do not even see it as harmful towards the brands, since they are established and money does not appear to be a problem. The counterfeit market, according to the sources found, has been growing steadily through the years; since manufacturers from China keep emerging and the sanction they pay if they get caught is not heavy enough to make them stop producing counterfeit goods, the profit they make is worth the risk. The chapter suggests the relevance and importance of alternative technology for tags and labels especially through the NFC technology to prevent counterfeit. The chapter relates with the anti-counterfeit software of BULLA to prevent counterfeit. BULLA could have a fighting chance, with this growth brands would want to take action when it comes to protecting their assets.

BULLA would probably work better on an international level, since in Portugal, manufacturers are not willing to spend extra funds to better or to protect their business; however, they do not suffer as greatly from this malady as bigger companies do. Bigger companies are willing to fund something as long as it has some appeal and some potential to increase their business. This means that the efforts regarding BULLA should be made in the direction of partnering up with bigger and solid companies; if not, the focus should at least be in selling our product to other technological companies and receiving a percentage of their profits; in this second option, the efforts on our part would be minimised.

MobilityNow is now an accredited entity under Portugal 2020. Its expertise and competency has been recognised in provision of mobility consultancy services. In addition, the company has been marketing at present the near-field communications and its multiple uses for the premium brand. Over the years, the strategies of reaching consumer through their anti-counterfeit products have been successful.

Author details

Amélia Maria Pinto da Cunha Brandão* and Mahesh Gadekar
Faculdade de Economia da Universidade do Porto, Portugal

*Address all correspondence to: ameliabrandao@fep.up.pt

References

[1] Krupka Z, Ozretic-Dosen D, Previsic J. Impact of perceived brand name origin on fashion brand's perceived luxury. Acta Polytechnica Hungarica. 2014

[2] Bian Q, Forsythe S. Purchase intention for luxury brands: A cross cultural comparison. Journal of Business Research. 2012;**65**(10):1443-1451

[3] Wang Y, Song Y. Counterfeiting: Friend or foe of luxury brands? An examination of Chinese consumers' attitudes toward counterfeit luxury brands. Journal of Global Marketing. 2013

[4] Lai KKY, Zaichkowsky JL. Brand imitation: Do the Chinese have different views? Asia Pacific Journal of Management. 1999:1-5

[5] Peng L, Wong AHK, Wan LCY. The effects of image congruence and self-monitoring on product evaluations: A comparison between genuine and counterfeit products. Journal of Global Marketing. 2012

[6] Ang SH, Cheng PS, Lim EAC, Tambyah SK. Spot the difference: Consumer responses towards counterfeits. Journal of Consumer Marketing. 2001

[7] Hieke S. Effects of counterfeits on the image of luxury brands: An empirical study from the customer perspective. Journal of Brand Management. 2010

[8] Yu C-C, Lin P-J, Chen C-S. How brand image, country of origin, and self-congruity influence internet users' purchase intention. Social Behavior and Personality: An International Journal. 2013

[9] Levy SJ. Symbols for sale. Brands, Consumers, Symbols and Research: Sidney J Levy on Marketing. 1999:203-212

[10] Poddar A, Foreman J, Banerjee SS, Ellen PS. Exploring the Robin Hood effect: Moral profiteering motives for purchasing counterfeit products. Journal of Business Research. 2012

[11] UNODC. The Globalisation of Crime: A Transnational Organized Crime Threat Assessment; 2010

[12] Johnson PE, Grazioli S, Jamal K, Berryman RG. Detecting deception: Adversarial problem solving in a low base-rate world. Cognitive Science. 2001

[13] Mavlanova T, Benbunan-Fich R. Counterfeit products on the internet: The role of seller-level and product-level information. International Journal of Electronic Commerce. 2011

[14] Ekman P. An argument for basic emotions. Cognition and Emotion. 1992

[15] Bloch PH, Bush RF, Campbell L. Consumer 'accomplices' in product counterfeiting: A demand-side investigation. Journal of Consumer Marketing. 1993

[16] OECD. Global Trade in Fake Goods Worth Nearly Half a Trillion Dollars a Year; OECD & EUIPO [Online]. 2016. Available from: http://www.oecd.org/industry/global-trade-in-fake-goods-worth-nearly-half-a-trillion-dollars-a-year.htm [Accessed: March 20, 2019]

[17] World Custom Organization. Global Operation Nets Tens of Thousands of Counterfeits [Online]. 2012. Available from: http://www.wcoomd.org/en/media/newsroom/2012/february/global-operation-nets-tens-of-thousands-of-counterfeits.aspx [Accessed: March 12, 2019]

[18] Wilcox K, Kim HM, Sen S. Why do consumers buy counterfeit luxury brands? Journal of Marketing Research. 2009

[19] Chen A, Peng N. Examining Chinese consumers' luxury hotel staying behavior. International Journal of Hospitality Management. 2014

[20] Dryden J. Counting the cost: The economic impacts of counterfeiting and piracy. In: Third Global Congress on Combating Counterfeiting and Piracy. 2007

[21] Havocscope. Counterfeit Black Markets; 2019

[22] Park H, Kim Y. The role of social network websites in the consumer–brand relationship. Journal of Retailing and Consumer Services. 2014

[23] Statista. Global Luxury Goods Industry-Statistics & Facts [Online]. 2019. Available from: https://www.statista.com/topics/1110/global-luxury-goods-industry/ [Accessed: March 16, 2019]

[24] Fionda AM, Moore CM. The anatomy of the luxury fashion brand. Journal of Brand Management. 2009

[25] The Fashion Law. The Counterfeit Report: The Big Business of Fakes [Online]. 2018. Available from: http://www.thefashionlaw.com/home/the-counterfeit-report-the-impact-on-the-fashion-industry [Accessed: March 20, 2019]

[26] The Counterfeit Report "Counterfeit" [Online]. 2019. Available from: https://thecounterfeitreport.com/ [Accessed: February 10, 2019]

[27] MarkMonitor. Brand Protection ROI [Online]. 2019. Available from: https://www.markmonitor.com/why-markmonitor/recover-revenue-with-brand-protection-roi

[28] US Customs and Border Protection. CBP Seizes $4.4 Million in Counterfeit Products in Puerto Rico and the USVI [Online]. 2019. Available from: https://www.cbp.gov/newsroom/local-media-release/cbp-seizes-44-million-counterfeit-products-puerto-rico-and-usvi

[29] Zheng J-H, Shen B, Chow P-S, Chiu C-H. The impact of the strategic advertising on luxury fashion brands with social influences. Mathematical Problems in Engineering. 2013

[30] Amaldoss W, Jain S. Pricing of conspicuous goods: A competitive analysis of social effects. Journal of Marketing Research. 2005

[31] Belk RW. Possessions and the extended self. Journal of Consumer Research. 2002

[32] Godey B, Lai C. Construction of international brand portfolios: Impact on local brands. The Journal of Product and Brand Management. 2011

[33] Han YJ, Nunes JC, Drèze X. Signaling status with luxury goods: The role of brand prominence. Journal of Marketing. 2010

[34] Olsen JE, Granzin KL. Using channels constructs to explain dealers' willingness to help manufactures combat counterfeiting. Journal of Business Research. 1993

[35] Radon A. Counterfeit luxury goods online: An investigation of consumer perceptions. International Journal of Marketing Studies. 2012

[36] Tom G, Garibaldi B, Zeng Y, Pilcher J. Consumer demand for counterfeit goods. Psychology and Marketing. 1998

[37] Delener N. International counterfeit marketing: Success without risk. Review of Business. 2000

[38] Shoham A, Ruvio A, Davidow M. (Un)ethical consumer behavior: Robin hoods or plain hoods? Journal of Consumer Marketing. 2008

[39] Xia L, Monroe KB, Cox JL. The price is unfair! A conceptual framework of price fairness perceptions. Journal of Marketing. 2004;**68**(4):1-15

[40] Cordell VV, Wongtada N, Kieschnick RL. Counterfeit purchase intentions: Role of lawfulness attitudes and product traits as determinants. Journal of Business Research. 1996

[41] Park H-J, Burns LD. Fashion orientation, credit card use, and compulsive buying. Journal of Consumer Marketing. 2005;**22**(3):135-141

[42] Xiao Lu P, Pras B. Profiling mass affluent luxury goods consumers in China: A psychographic approach. Thunderbird International Business Review. 2011

[43] Taylor T. What is Content Marketing? 2014. Available from: http://www.asperoi.com/blog/what-is-content-marketing/ [accessed in January 28 th 2015]

[44] Pulizzi J. Content Marketing how-to: The objective approach. Communication World. 2013;**11**:19-22. Available from: http://cwarchives.iabc.com/communicationworld/november_2013?pg=19#pg19 [accessed in January 28 th 2015]

Chapter 5

The RFId Technology for Monitoring the Supply Chain and for Fighting against Counterfeiting: A Fashion Company Case Study

Erica Varese and Anna Claudia Pellicelli

Abstract

The purpose of this chapter, after a brief literature review, is to analyse how the RFId technologies applied by an Italian fashion firm, Oscalito, contribute to monitoring the supply chain and are a useful tool to fight against counterfeiting, enhancing the Made in Italy. In the textile sector, characterised by a short and constantly evolving production cycle, the RFId technology has enormous potential. According to Regulation (EU) No 1007/2011 on textile fibre names and related labelling and marking of the fibre composition of textile products, consumer protection requires transparent and consistent trade rules, including as regards, indications of origin. When such indications are used, they should enable consumers to be fully aware of the origin of the products they purchase, so as to protect them against fraudulent, inaccurate or misleading claims of origin. In this context, RFId technology has emerged as a valid support for the company not only to monitor the supply chain, especially with reference to inventory management, waste disposal, logistics and transport, but also to protect the Italian origin of production. This study also has some limitations, typical of the applied methodology.

Keywords: RFId, supply chain, counterfeiting, fashion company, label, case study

1. Introduction

The Textile Regulation (EU) No 1007/2011 on textile fibre names and related labelling and marking of the fibre composition of textile products (hereafter, Regulation) was adopted in September 2011 and became applicable on 8 May 2012. It repealed and replaced the previous Textile Directives.

By it, the EU legislator intended to eliminate potential obstacles to a good functioning of the internal market, caused by divergent rules in the member states.

The enforcement of this Regulation thus aims at standardising textile fibre names as well as terms appearing on labels, markings and documents accompanying textile products in the various production, transformation and distribution cycles.

With the double objective of offering consumers accurate information and of improving the internal market, it also sets rules on labelling or marking of textile products containing non-textile parts of animal origin, and regulates modalities for determining the fibre composition of textile products.

A textile product is defined as "*any raw, semi-worked, worked, semi-manufactured, manufactured, semi-made-up or made-up product which is exclusively composed of textile fibres, regardless of the mixing or assembly process employed*" (Regulation, art. 3).

The burden of labelling lies on all producers, importers or distributors of textile products, from raw materials to the finished product.

At the time of introducing a product into the market, the producer ensures the presence of the label or marking and the accuracy of the information therein displayed. If the producer is not established in the Union, the importer guarantees the above. A distributor is considered producer to the scopes of the Regulation in case he introduces a product into the market with his name or factory mark, labels it or modifies its contents. When making a textile product available on the market, the distributor guarantees that it is provided with the appropriate label or marking (Regulation, art. 15).

The Regulation does not apply to textile products contracted out to persons working in their own homes or to independent firms that carry out work from materials supplied without the property therein being transferred for consideration or to textile products tailor-made by self-employed tailors.

Furthermore, the following do not require a label: (i) textile products not intended for final consumers; (ii) textile products which, under customs control, are in transit because they are intended for non-EU markets; (iii) textile products temporarily imported for processing; (iv) textile products intended for sale in non-EU countries, for which the rules of the destination country need to be respected.

On top of mandatory indications, the legislator allows operators to characterise their production by voluntary information to be applied on labels or markings. In recent years, smart labels [1] are spreading out more and more in the textile sector also.

The objective of this chapter is to present, among indications which can voluntarily be added to labels, the application by Oscalito of RFId tags on finished items of clothing, in order to guarantee control on supply chain, on Made in Italy and, in general, on the high-quality characteristics of production.

This research fills a gap in literature: to the authors' knowledge, this is the first paper to present a case study on an Italian company applying RFId technologies to control the supply chain and to protect the Italian origin of production.

This chapter is divided into five sections: Section 2 presents a brief literature review on RFId technologies in the textile sector for monitoring the supply chain and for fighting against counterfeiting; Section 3 shows the research methodology; Section 4 offers considerations regarding the case study, Oscalito; in Section 5, the final conclusions are summarized.

2. Literature review

RFId technology uses a radio frequency to identify, detect and locate objects [2]. In a nutshell, these systems are based on remote reading of the information displayed by a specific label (the RFId tag), activated by a special reader. Thus, it is possible, through magnetic impulses, to codify the data contained in the tag accompanying the product throughout the entire production process [3]. Passive RFId tags do not have own power supply; therefore, the chips are activated by the power received from the antenna of the reader.

Literature analysis primarily brought up two main uses of RFId tags in the textile industry: (i) for monitoring supply chains in general and (ii) as tools for fighting against counterfeiting.

2.1 RFId technologies for monitoring the supply chain

Numerous publications describe potential benefits of the use of RFId tags in the supply chain. For example, see [4–11]. Few authors succeed in quantifying the benefits deriving from the application of this technology, because of the relatively scarce concrete applications [12].

In recent years, RFId tags have risen great interest, and some authors [12] believe that they will substitute bar codes, an automatic identification technology that has been characterising retail sales for decades [13].

They are used for monitoring supply chains especially with reference to inventory management, waste disposal, logistics and transport [14].

In the textile industry, characterised by a short and continuously evolving production cycle, RFId technology enjoys enormous potential. Brands such as Prada, Tesco, Wal-Mart, Benetton and, recently, Zara [15] are studying possibilities for its implementation [16]. The positive impact of these tags has been analysed by [17], who presented the highly satisfactory results obtained along the supply chain by applying RFId tags to clothing items of a US company. In Ref. [18], on the other hand, in a Hong Kong company, it has been noted that a resource allocation system based on RFId tags can ensure more efficient processes than those obtainable by traditional techniques.

The Italian textile and clothing industry has for years been struck by strong competition from emerging economies: this situation led some authors [19] to verify the existence of traceability initiatives in order to obtain competitive advantages. The study analysed the 'Traccia' project for dissemination of the 'traceability of textile products', carried out also by using RFId tags. It was brought into evidence that various traceability models might have come up, supported by public or private certification systems. Even though the textile industry is considered one of the most indicated for the application of this technology [20], the authors lament that operators do not yet fully perceive its advantages.

The main responsibility for the delay in implementing these tags, on the Italian as well as the international level, seems to be upon difficulties in aligning objectives and strategies along the supply chain: if this were not to happen, RFId might hinder instead of favouring integration of the various processes [21]. In Ref. [22], finally, a possible application of RFId to the shopping context is suggested: by guiding consumers in their purchases, added value could be created for firms.

Furthermore, numerous other interesting studies concern applications of RFId tags to the textile industry (e.g., see [22–26]).

2.2 RFId technologies for fighting against counterfeiting

In the twenty-first 'Whereas' of the Regulation, the EU legislator states that the textile industry is hit by the phenomenon of counterfeiting and that this raises problems in terms of consumer protection and information. The legislator encourages member states to devote particular attention to the enforcement of EU horizontal legislation and of measures concerning counterfeited products in the textile industry, such as Regulation (EU) No 608/2013 concerning customs enforcement of intellectual property rights and repealing Council Regulation (EC) No 1383/2003.

Technologies enabling contrast counterfeiting can be classified into four groups [27–29]: (i) holograms and filigrees (manifest technologies), (ii) safety inks and invisible printing (hidden technologies), (iii) chemical tags and (iv) bar codes and RFId ('track-and-trace' technologies).

On the international scale, the proposal and analysis of RFId tags in order to fight against counterfeiting have been studied by many authors (e.g., see [30–40]).

In Ref. [41], it is suggested that consumers should use personal mobile devices (with RFId reader) in order to obtain information on products they are about to purchase and verify, in particular, their authenticity, while in Ref. [42], it is proposed to integrate RFId tags through the innovative anti-counterfeiting 'TagPrint' system, using COTS RFID tags and readers. This system is characterised by low-cost and offline genuineness validation utilizing passive tags. These three purposes are achieved "*by leveraging a few of federated tags' fingerprints and geometric relationships*". In TagPrint, a new kind of fingerprint is utilised, called phase fingerprint, "*extracted from the phase value of the backscattered signal, provided by the COTS RFID readers*". To further solve the separation challenge, a geometric solution is developed to validate the genuineness. TagPrint, using COTS RFID devices, may increase the inviolability of RFId tags.

Concerning the textile industry, Ref. [43] presents a practical application of the RFId technology to the fashion sector of an Italian firm: it is argued that the implementation of this technology as to the two most imitated lines of production has enabled to limit the counterfeiting phenomenon and to improve logistics.

Still with a view to the textile industry, thanks to using RFId technology, Refs. [44, 45] propose a system called 'electronic-pedigree' (e-pedigree), which enables to verify single elements, identify missing objects and foresee the status of the products wherever they are located within the supply chain. In 2015, an algorithm named 'tag data processing and synchronization—TDPS' was presented, which makes it possible to develop an e-pedigree [29]. To date, the studies of these authors however still miss practical applications.

On top of eliminating sales of counterfeited goods, the application of RFId tags on items of clothing enables to hamper organised crime, by rapidly identifying and reacting to its illicit strategies, which are constantly evolving [46].

Further interesting cases are mentioned in Refs [47, 44].

3. Methodology

With the purpose to achieve the aim of this research, the following hypothesis has been developed:

H1. *In the fashion industry there is an increasing need to monitor the textile supply chain; in Italy, furthermore, it is necessary to protect the textile production against counterfeiting. Since RFId technologies are considered a strategic tool in many sectors (for example, in traceability of food, animals, and people thanks to electronic passports), the Italian textile industry may be very interested in approaching them*.

The research methodology was structured as follows: the first stage consisted in a review of existing literature, focused on RFId technologies in the textile sector for monitoring the supply chain and for fighting against counterfeiting; the second stage consisted in applying a qualitative case study methodology helping to explore this phenomenon within its context [48, 49].

According to Ref. [50], the choice of this methodology is justified by the need to answer 'how' and 'why', as well as by the fact that authors cannot manipulate the behaviour of those involved in the study, and by the fact that the research focuses on a contemporary phenomenon [51].

We feel that it would be impossible to gain a true picture of the chances for adopting an RFId technology in the fashion industry without considering the context in which it may be developed and used.

In fact, the relationship between RFId technology on the one side, and its implementation for monitoring the supply chain and avoiding counterfeiting on the other side, seems crucial to us.

We chose this case, Oscalito, because it is quite unique [52, 50, 53].

Oscalito undoubtedly shows the above-mentioned characteristics, because it has been one of the first Italian firms to implement voluntary RFId labels.

According to Ref. [54], this essay uses a wide range of sources of information in order to develop and analyse the case study. In the interest of data triangulation, we observed directly, analysed company documents and made interviews.

Direct observation was conducted at the company premises in 2018, so as to catch the reality and analyse events in real time: we enjoyed the opportunity to observe several meetings. We are conscious of the weaknesses of such observation: time-consuming; selectivity (might miss facts), reflexivity (observer's presence might cause change) and costs (observers need time)—[55, 51].

On these occasions, we asked to be granted access to company documents in order to better understand the firm and to increase our knowledge about the enterprise, especially concerning the RFId label.

We had the opportunity to analyse memoranda, study reports, etc. The validity of these documents was carefully reviewed so as to avoid incorrect data being analysed. We spent almost a week collecting data emerging from this documentation. Further information was collected from the Oscalito website.

In order to capture different dimensions of the same phenomenon, we interviewed the managing director and other people in the company (semi-structured interview) so as to clarify some important topics [54, 56, 57]. Each interview lasted for approximately 1 hour and was conducted by both of us. With a view to reducing the subjectivity of data interpretation, on permission by the interviewee [51], the interviews were recorded and later transcribed.

We autonomously analysed all data obtained by direct observation, company documentation and interviews, and we finally compared our individual interpretation of the results.

We did a triangulation of data sources (data triangulation) [58].

According to the Ying categorisation of case studies, this is a 'descriptive' one: this type of case study is used to describe a 'phenomenon and the real-life context in which it occurred' [50, 49].

4. Case study: Oscalito

In order to better understand the practical applications smart labels can have in the textile industry, the CEO of Oscalito, Dr. Dario Casalini, was contacted and interviewed.

The Oscalito brand (acronym for Osvaldo Casalini Lino Torino) is produced by Maglificio Po, a textile company founded in Turin in 1936 by two brothers, Osvaldo and Lino Casalini. Initially, clothing lines of underwear and fashion knitwear for men, women and children were created, using high-quality natural fibres. Basically, tubular fabric (without stitching) was produced, using circular machines. Lino's sons, Arrigo and Andrea, later joined the firm, extending the product range to fashion clothes and gaining success even on foreign markets. Export became fundamental for the business (so fundamental that nowadays 70% of the production is exported), and over the years, the women's fashion line acquired a central role in Oscalito's offer. In 2014, Dario Casalini joined the firm; as third generation of the family, he gradually took the lead of the company, renewing the brand and making it grow on the international scale, but always in the sign of continuity. In fact, Oscalito has always kept the entire production chain in Turin: from yarn production to finished garment. It has been among the first firms in the textile industry to introduce RFId labels on individual items of clothing in order to ensure production chain traceability.

Since 2012, Oscalito has been using RFId tags, which allow to trace the entire chain: fabric weaving, finishing and cutting, tailoring, quality control and logistics. This makes absolute control on quality flaws possible and guarantees Made in Italy. Such instrument enables to enact the intent of the EU Legislator, who, at the twentieth "whereas" of the Regulation, with the aim to protect consumers, requests transparent and coherent commercial rules, also concerning indications of origin.

These indications, when available, should enable consumers to be fully informed on the origin of the products they purchase, so as to protect them from indications of origin which are fraudulent, inaccurate or misleading.

Oscalito's production chain is managed through an ERP which uses bar code technology in the first two processing phases [(i) fabric weaving and finishing; (ii) fabric and lace cutting], while during the further four phases [(iii) tailoring the completed garments; (iv) finishing the completed garments; (v) ironing and packaging; (vi) logistics and warehousing], the information contained in the bar code is poured out into the RFId tag and therefore individualised for every single piece of clothing.

The implementation of RFId technology in the above production phases took place, as mentioned, starting from May 2012; it foresees continuous printing of the cards with RFId tags even in case part of the production cycle is carried out by third parties (**Figure 1**).

The RFId tag is applied following quality control which is made on every garment, and it enables to manage the phases of repair of items which are faulty but can be recuperated. Applying RFId tags allows to trace all production phases and operators which came in the finished garment in case it was returned, bringing to evidence any shortcoming or organizational problem in one of the production phases.

The implementation of RFId tags automatized all logistic functions, such as entrance into the warehouse with production notes and inventory functions by

Figure 1.
RFId Oscalito. Particular of the 'maintenance' card with RFId tag associated to the guarantee of Made in Italy product. Source: Oscalito.

reader after passing through the tunnel (**Figure 2**) or by shelf reader scanner, as well as deliveries to customers.

Oscalito does not have an automatized warehouse; therefore, customer order preparation is done by manual shelf picking. RFId tags are fundamental in controlling picking accuracy: by printing the labels, first of all the true contents of every single box are brought into evidence, and then a further control of the complete order compared to the picking list (i.e., the customer's order) is carried out. Finally, packing lists for every single item and for the total delivery are printed (**Figure 3**), as well as the documents needed by the forwarder.

The results obtained by Oscalito may be summarized as follows:

- Complete traceability of the chain: from yarn to finished garment, in all production phases, with evidence of the operators who came in during the various processing steps

- Analysis of defects and errors in the various production phases

- Management of finished garments repair

- Warehouse and inventory management

- Delivery management: cancelling of human mistakes in picking up; automatic printing of labels on the boxes; note and packing list printing for every single item

With reference to countering counterfeiting, RFId system is ideal since each RFID tag generated has a 22-character alphanumeric code that is unique and not reproducible. It contains product production data that can be extracted via the ERP system that generated it and which is the only one that allows verification of the authenticity and originality of the product. The final consumer can verify the authenticity of the product by entering the 22-character alphanumeric code in a form made available by the manufacturer on its own mobile app or website.

The manufacturer can easily create his own mobile app or website where the final consumer can verify the authenticity of the product by entering the 22-character alphanumeric code of the RFID in a form made available by the manufacturer. By promoting this autonomous verification system by the final customer, a widespread control is carried out on the authenticity of the product without direct costs for the manufacturing company.

Figure 2.
RFId tunnel with exchanging direction motor conveyor for warehouse entry and delivery exit. Source: Oscalito.

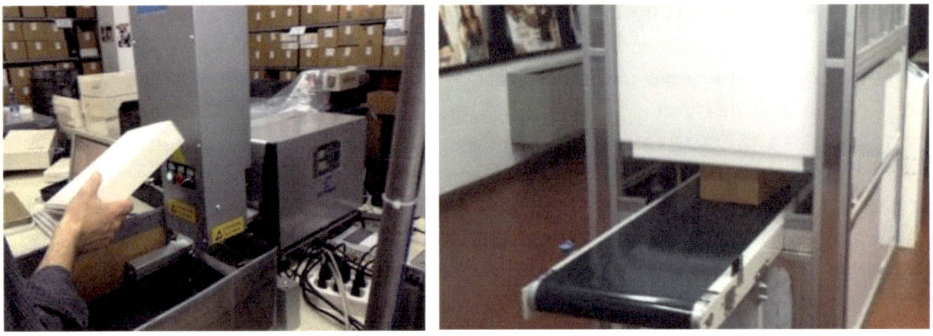

Figure 3.
RFId reading post, Box label printing and application. Source: Oscalito.

As its future objective, Oscalito aims at extending RFId technology to all fabric weaving, finishing and cutting production phases and to warehouse management by customer, with pre-allocation of garments.

5. Conclusions

The implementation of a production chain traceability system by RFId tags and their readers (printers for initializing cards with RFId tags; tunnels to read warehouse entries and exits; printers for labelling boxes when dispatching) implied an investment of about 70,000.00 euros, increased by costs for the adaptation of the informatics management system and by current expenditure concerning periodical RFId tag purchase.

Many benefits were obtained: from production chain management with traceability of garments being repaired, to automatic management of the general warehouse and of consignment accounts with some customers, to management of commercial or defectiveness returns by customers and final consumers who purchase through Oscalito's e-commerce portal, on top of annulment of mistakes in the dispatching phase, eliminating claims on contents of the dispatched pieces, which used to concern about 100 pieces a year.

Finally, the implementation of the production chain traceability system by RFId tags has enabled to obtain important supply chain certifications, especially the Italian Identity certification issued by Italcheck.

Even if, according to Refs. [59, 60], common criticisms of the case study method are that it lacks rigour and that the dependence on a single case exploration makes it difficult to reach a generalising conclusion, the authors believe that through the Oscalito case study, they have been able to describe their views on a relevant innovative reality.

We realise that this research has some limitations due to the applied methodology: we have adopted a qualitative method for a single case study; the findings of the study are based on the first results of prospected deeper research; and further interviews would be required in order to understand a general fashion industry perception of this kind of technology.

For this reason, we are planning to carry out further studies on the application of RFId in the fashion industry.

Acknowledgements

The authors would like to thank the editor and the anonymous referees for their comments and helpful suggestions and Dr. Dario Casalini for his support and critical reading of the chapter.

Conflict of interest

The authors have no conflict of interest.

Other declarations

Both authors contributed equally to the chapter.

Author details

Erica Varese* and Anna Claudia Pellicelli
Department of Management, University of Torino, Torino, Italy

*Address all correspondence to: erica.varese@unito.it

References

[1] Scholz-Reiter B, Teucke M, Özsahin ME, Sowade S. Smart label-supported autonomous supply chain control in the apparel industry. In: Proceedings of the 5th International Congress on Logistics and SCM Systems (ICLS 2009); 2-5 June 2009; Seoul, Korea. pp. 44-52

[2] Varese E, Buffagni S, Percivale F. Application of RFId technology to the agro-industrial sector: Analysis of some case studies. Journal of Commodity Science, Technology and Quality. 2008;**47**(1-4):171-190

[3] D'Ascenzo F. RFId: Feature of technologies and applications in a concrete experimental environment. Journal of Commodity Science, Technology and Quality. 2005;**44**(2-4):99-112

[4] Kelly EP, Erickson GS. RFID tags: Commercial applications vs. privacy rights. Industrial Management & Data Systems. 2005;**105**(6):703-713. DOI: 10.1108/02635570510606950

[5] Asif Z, Mandviwalla M. Integrating the supply chain with RFID: A technical and business analysis. Communications of the Association for Information Systems. 2005;**15**(24):1-57. DOI: 10.17705/1CAIS.01524

[6] Prater E, Frazier G, Reyes PM. Future impacts of RFID on e-supply chains in grocery retailing. Supply Chain Management: An International Journal. 2005;**10**(2):134-142. DOI: 10.1108/13598540510589205

[7] Delen D, Hardgrave BC, Sharda R. RFID for better supply-chain management through enhanced information visibility. Production and Operations Management. 2007;**16**(5):613-624. DOI: 10.1111/j.1937-5956.2007.tb00284.x

[8] Gaukler GM, Seifert RW. Applications of RFID in supply chains. In: Jung H, Chen FF, Bongju J, editors. Trends in Supply Chain Design and Management: Technologies and Methodologies. Heidelberg: Springer; 2007. pp. 29-48. DOI: 10.1007/978-1-84628-607-0_2

[9] Visich JK, Li S, Khumawala BM, Reyes PM. Empirical evidence of RFID impacts on supply chain performance. International Journal of Operations & Production Management. 2009;**29**(12):1290-1315. DOI: 10.1108/01443570911006009

[10] Zhu X, Mukhopadhyay SK, Kurata H. A review of RFID technology and its managerial applications in different industries. Journal of Engineering and Technology Management. 2012;**29**(1):152-167. DOI: 10.1016/j.jengtecman.2011.09.011

[11] Bottani E, Bertolini M, Rizzi A, Romagnoli G. Monitoring on-shelf availability, out-of-stock and product freshness through RFID in the fresh food supply chain. International Journal of RF Technologies. 2017;**8**(1-2):33-55. DOI: 10.3233/RFT-171780

[12] Al-Kassab J, Blome P, Wolfram G, Thiesse F, Fleisch E. RFID in the apparel retail industry: A case study from Galeria Kaufhof. In: Unique Radio Innovation for the 21st Century. Berlin Heidelberg: Springer; 2011. pp. 281-308. DOI: 10.1007/978-3-642-03462-6_13

[13] Manthou V, Vlachopoulou M. Bar-code technology for inventory and marketing management systems: A model for its development and implementation. International Journal of Production Economics. 2001;**71**(1):157-164. DOI: 10.1016/S0925-5273(00)00115-8

[14] Varese E, Beltramo R, Buffagni S, Capello F, Toja M, Gallino G. Benefits of real-time monitoring logistics in

agribusiness. Case study: See your box service based on industrial internet of things (IIoT) technology. In: Food & Heritage: Sostenibilità Economico-Aziendale e Valorizzazione del Territorio. Torino, IT: Giappichelli; 2015. pp. 131-142

[15] Yip AC, Huang M. Strategic values of technology-driven innovation in inventory management: A case study of Zara's RFID implementation. International Journal of Inventory Research. 2016;**3**(4):318-336. DOI: 10.1504/IJIR.2016.082326

[16] Nayak R, Singh A, Padhye R, Wang L. RFID in textile and clothing manufacturing: Technology and challenges. Fashion and Textiles. 2015;**2**(1):9. DOI: 10.1186/s40691-015-0034-9

[17] Singh SP, McCartney M, Singh J, Clarke R. RFID research and testing for packages of apparel, consumer goods and fresh produce in the retail distribution environment. Packaging Technology and Science. 2008;**21**(2):91-102. DOI: 10.1002/pts.782

[18] Lee CKH, Choy KL, Ho GT, Law KMY. A RFID-based resource allocation system for garment manufacturing. Expert Systems with Applications. 2013;**40**(2):784-799. DOI: 10.1016/j.eswa.2012.08.033

[19] Pigni F, Crave S, Aurelio R. Traceability in the textile and clothing industry: Issues and implications for RFId adoption. In: The 2nd Mediterranean Conference on Information Systems. Venezia, IT: MCIS; 2007

[20] Barburski M, Czekalski B, Snycerski M. RFID technology in the textile industry. Autex Research Journal. 2008;**8**(3):92-96

[21] Ruile H, Wunderlin P. RFID application within product life cycle of industrial textile & apparels. In: Proceedings of RFID SysTech 2011 7th European Workshop on Smart Objects: Systems, Technologies and Applications; 17-18 May 2011; Dresden, Germany. pp. 1-11

[22] Chen CC. RFID-based intelligent shopping environment: A comprehensive evaluation framework with neural computing approach. Neural Computing and Applications. 2014;**25**:1685-1697. DOI: 10.1007/s00521-014-1652-7

[23] Keung Kwok S, Wu KK. RFID-based intra-supply chain in textile industry. Industrial Management & Data Systems. 2009;**109**(9):1166-1178. DOI: 10.1108/02635570911002252

[24] Ngai EWT, Peng S, Alexander P, Moon KK. Decision support and intelligent systems in the textile and apparel supply chain: An academic review of research articles. Expert Systems with Applications. 2014;**41**(1):81-91. DOI: 10.1016/j.eswa.2013.07.013

[25] Rizzi A, Romagnoli G, Thiesse F. A new framework for RFID use cases in fashion and apparel retailing. International Journal of RF Technologies. 2016;**7**(2-3):105-129. DOI: 10.3233/RFT-150075

[26] Oner M, Ustundag A, Budak A. An RFID-based tracking system for denim production processes. The International Journal of Advanced Manufacturing Technology. 2016; **1**(90):591-604. DOI: 10.1007/s00170-016-9385-7

[27] Bansal D, Malla S, Gudala K, Tiwari P. Anti-counterfeit technologies: A pharmaceutical industry perspective. Scientia Pharmaceutica. 2013;**81**(1):1-13. DOI: 10.3797/scipharm.1202-03

[28] Li L. Technology designed to combat fakes in the global supply chain. Business Horizons. 2013;**56**(2):167-177. DOI: 10.1016/j.bushor.2012.11.010

[29] Choi SH, Yang B, Cheung HH, Yang YX. RFID tag data processing in manufacturing for track-and-trace anti-counterfeiting. Computers in Industry. 2015;**68**:148-161. DOI: 10.1016/j.compind.2015.01.004

[30] Koh R, Schuster EW, Chackrabarti I, Bellman A. White Paper: Securing the Pharmaceutical Supply Chain. Auto-ID Centre; 2003

[31] Jansen-Vullers MH, van Dorp CA, Beulens AJ. Managing traceability information in manufacture. International Journal of Information Management. 2003;**23**(5):395-413. DOI: 10.1016/S0268-4012(03)00066-5

[32] Staake T, Thiesse F, Fleisch E. Extending the EPC network: The potential of RFID in anti-counterfeiting. In: Proceedings of the 2005 ACM Symposium on Applied Computing; 13-17 March 2005; Santa Fe, New Mexico. 2005. pp. 1607-1612. DOI: 10.1145/1066677.1067041

[33] Tuyls P, Batina L. RFID-tags for anti-counterfeiting. In: Pointcheval D, editor. Topics in Cryptology—CT-RSA 2006. Lecture Notes in Computer Science. Vol. 3860. Berlin, Heidelberg: Springer; 2006. pp. 115-131. DOI: 10.1007/11605805_8

[34] Cole PH, Ranasinghe DC. Networked Rfid Systems & Lightweight Cryptography. London, UK: Springer-Verlag; 2007

[35] Staake T, Michahelles F, Fleisch E, Williams JR, Min H, Cole PH, et al. Anti-counterfeiting and supply chain security. In: Cole P, Ranasinghe D, editors. Networked RFID Systems and Lightweight Cryptography. Berlin, Heidelberg: Springer; 2008. pp. 33-43. DOI: 10.1007/978-3-540-71641-9_2

[36] Choi SH, Poon CH. An RFID-based anti-counterfeiting system. International Journal of Computer Science. 2008;**35**(1):1-12

[37] Lehtonen M, Staake T, Michahelles F, Fleisch E. From identification to authentication: A review of RFID product authentication techniques. In: Networked RFID Systems and Lightweight Cryptography, Part II. 2008. pp. 169-187. DOI: 10.1007/978-3-540-71641-9_9

[38] Yan B, Huang G. Application of RFID and Internet of Things in monitoring and anti-counterfeiting for products. In: Proceedings of the International Seminar on Business and Information Management; 19-19 December 2008; Wuhan, China. 2009. pp. 392-395. DOI: 10.1109/ISBIM.2008.196

[39] Ilic A, Lehtonen M, Michahelles F, Fleisch E. Synchronized secrets approach for RFID-enabled anti-counterfeiting. In: Demo at Internet of Things Conference. 2008

[40] Cheung HH, Choi S. Implementation issues in RFID- based anti-counterfeiting systems. Computers in Industry. 2011;**62**:708-718. DOI: 10.1016/j.compind.2011.04.001

[41] Kim J, Kim H. Anti-counterfeiting solution employing mobile RFID environment. International Journal of Electrical, Computer, Energetic, Electronic and Communication Engineering. 2007;**1**(8):1200-1204

[42] Yang L, Peng P, Dang F, Wang C, Li XY, Liu Y. Anti-counterfeiting via federated RFID tags' fingerprints and geometric relationships. In: 2015 IEEE Conference on Computer Communications (INFOCOM); 26 April-1 May 2015; Kowloon, Hong Kong. New York: IEEE; 2015. pp. 1966-1974. DOI: 10.1109/INFOCOM.2015.7218580

[43] Pigni F, Ravarini A, Astuti S, Buonanno G, Mari L. RFID in The Fashion Industry: Exploratory Case Study And Implications.

Conference Paper. 2006. Available from: https://s3.amazonaws.com/academia.edu.documents/40607958/RFID_IN_THE_FASHION_INDUSTRY_EXPLORATORY20151203-18169-19zl3rc.pdf?AWSAccessKeyId=AKIAIWOWYYGZ2Y53UL3A&Expires=1551545601&Signature=Wy8NKSQj9N%2FDW6QnWflC1u%2BaZbc%3D&response-content-disposition=inline%3B%20filename%3DRFId_in_the_Fashion_Industry_Exploratory.pdf [Accessed: March 17, 2019]

[44] Choi SH, Yang B, Cheung HH, Yang YX. Data management of RFID-based track-and-trace anti-counterfeiting in apparel supply chain. In: 8th International Conference on Internet Technology and Secured Transactions (ICITST); 9-12 December 2013; London, UK. New York: IEEE; 2013. pp. 265-269. DOI: 10.1109/ICITST.2013.6750203

[45] Choi SH, Cheung HH, Yang B, Yang XY. Implementation issues in RFID-based anti-counterfeiting for apparel supply chain. International Journal of Intelligent Computing Research. 2014;**5**(1/2):379-389

[46] Liard M. RFID Item-Level Tagging In Fashion Apparel And Footwear. ABI Research, Oyster Bay, NY Q4. 2009. Available from: https://www.tycoifs.co.uk/documents/uploads/brochure/CR-ADT-101Final_RFID.pdf [Accessed: March 17, 2019]

[47] Wong KHM, Hui PCL, Chan ACK. Cryptography and authentication on RFID passive tags for apparel products. Computers in Industry. 2009;**57**:342-349. DOI: 10.1016/j.compind.2005.09.002

[48] Yin RK. Case Study Research: Design and Methods. Beverly Hills, CA: Sage; 1984

[49] Baxter P, Jack S. Qualitative case study methodology: Study design and implementation for novice researchers. The Qualitative Report. 2008;**13**(4):544-559

[50] Yin RK. Case Study Research: Design and Methods. 3rd ed. Thousand Oaks, CA: Sage; 2003

[51] Yin RK. Case Study Research: Design and Methods. 5th ed. Thousand Oaks, CA: Sage; 2013

[52] Stake RE. The Art of Case Study Research. Thousand Oaks, CA: Sage; 1995

[53] Noor KBM. Case study: A strategic research methodology. American Journal of Applied Sciences. 2008;**5**(11):1602-1604

[54] Eisenhardt KM. Building theories from case study research. Academy of Management Review. 1989;**14**(4):532-550

[55] Tellis WM. Application of a case study methodology. The Qualitative Report. 1009;**3**(3):1-19

[56] Corbetta P. La Ricerca Sociale: Metodologie e Tecniche, le Tecniche Qualitative. Bologna, IT: Il Mulino; 2003

[57] Alvesson M. Methodology for close up studies-struggling with closeness and closure. Higher Education. 2003;**46**(2):167-193

[58] Patton MQ. Qualitative Research & Evaluation Methods. Thousand Oaks, CA: Sage Publication; 2002

[59] Tellis WM. Introduction to case study. The Qualitative Report. 1997;**3**(2):1-14

[60] Zainal Z. Case study as a research method. Jurnal Kemanusiaan. 2007;**9**:1-6

Chapter 6

How to Make Fabrics Talk Environment: The Scatol8 per la Sostenibilità Way

Riccardo Beltramo, Jacopo Beltramo, Paolo Cantore and Annalisa Romani

Abstract

The mission of *Scatol8 per la Sostenibilità srl* (in English Scatol8 for Sustainability) is the realization of products and services to communicate the theme of sustainable development. The economic viability of the entrepreneurial initiative is ensured by three lines of activity: (1) the realization of personalized management systems, (2) carrying out environmental education activities, and (3) design and manufacture of products. Scatol8's products are unique design objects—therefore exclusive—crafted with materials and techniques derived from scientific research. Osmosis between new researches in the fields of business management, chemistry, environment, and the Scatol8 company leads to the manufacture of products that incorporate recent innovations and transfer the knowledge content to those who can afford them, meaning those who have culture enough to understand and appreciate them. It has been said: luxury is experiential. This is the concept we aim for with our products. The following chapter describes the qualities of some sustainable luxury products and deepens the ways of thinking, designing, and making a line of garments. These garments incorporate electronic devices with natural fabrics: cotton, hemp, and eco-leather. This is an example of upcycling as a result of technological innovation, which is then analyzed by the design thinking.

Keywords: fashion industry, product innovation, mass customization, smart textiles, interactive apparels, eco-fashion, natural fibers, natural dyeing, business models, fashion designers, wearable

1. The Scatol8

Scatol8 per la Sostenibilità srl (briefly, Scatol8 srl) is an academic spin-off of the University of Turin and an innovative start-up. The company, founded in 2016, is the natural consequence to extensive research programs at the national and international level on the relationship between the use and transformation of resources by economic organizations and the quality of the environment, conducted by the Commodity Science Area of the Department of Management, University of Torino. It works in the field of clean technologies and environmental monitoring. Its activity concerns the development, production, and marketing of innovative products and services with a high technological value and more specifically the development, production, and marketing of integrated electronic systems and networks for the collection of variables—in particular environmental—and analysis of data collected through intelligent dashboard, primarily intended for agricultural, commercial, and industrial activities. Product innovativeness is due to the development of specific algorithms that allow the application of methodologies (business intelligence and organizational intelligence) traditionally reserved for corporate information to environmental and energy data. These products have a high technological value because of the specifically developed software which makes them modular and highly customizable; in particular it allows both a simple connection to preexisting environmental management systems and complete technological integration with sensor and actuator networks (Internet of Things).[1]

1.1 Fundamental parts of Scatol8

Scatol8 is a sensor platform for environmental and management variable monitoring. The aim is to measure all those variables that are relevant for the interaction between an economical organization and the environment. Examples of these variables are electrical consumption, gas emission in air, waste production, etc. The data collected through the sensor platform are sent to a remote server (see **Figure 1**) that offers the opportunity to check the real-time values as well as to look at the historical series.

The data collected in the remote database are not only available from a web interface that is easy to use, but values can be elaborated, aggregated, and compared also among different networks. Scatol8 is mainly focused upon the environmental sustainability, and it's based on these criteria:

Accessibility: Hardware and software are fully based on open-source technologies and software in view of cost containment, openness, and ease of access, even for training and academic purposes.

Environmental compatibility: When possible, all electronic devices are placed in recycled containers, coming mainly from consumer goods and electronics industry, transformed and adapted to their new function, or in containers made of wood (a renewable resource) or even cardboard.

Modularity: The system is constituted from time to time, according to the requirements and specifications of each application.

Dissemination: Scatol8 is not only a product but also an initiative to spread knowledge, which aims to involve young people in the creation of technology (and not only in its use). Along with the hardware output of the company, there is a continuously updated line of information about observed variables, sustainability, and purposes.

We can logically divide the monitored variables in three main groups as shown in **Table 1**: the environmental, the managerial, and the biomedical parameters. The web application showing the measured values is called Crusc8 (dashboard). Many indicators, one per variable, compose it. **Figure 2** shows an example of Crusc8 with nine variables in the same network (on the left). For each indicator, it is possible to look at his past trend as shown on the right. These examples are taken from a monitored school; in particular the graph represents the measured temperature in a classroom where the data were collected every 15 minutes.

[1] *Scatol8 per la Sostenibilità srl*, established in 2016, draws back from various national and international research projects conducted within the Department of Management of the University of Turin. For any additional information, please visit http://scatol 8.net.

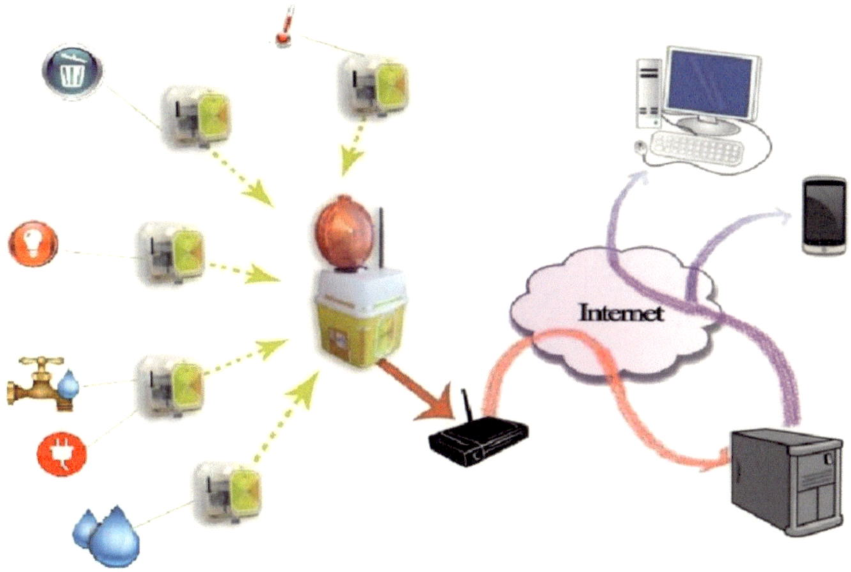

Figure 1.
An example of Scatol8 sensor network.

Environmental parameter
Acceleration
Power consumption
Wind direction
Distance (e.g., snow height, tank liquid quantities, etc.)
Liquid flow
Gas (presence of smoke, benzene, carbon dioxide, LPG, propane, hydrogen, oxygen, methane, carbon monoxide)
Illuminance
Mass (e.g., waste production)
Movement (e.g., intrusion, number of pieces, etc.)
Oxidation-reduction potential
pH
Rain
Atmospheric pressure
Radioactivity (decays α β γ)
Noise
Liquid temperature
Soil temperature
Air temperature
Soil humidity
Air humidity
Wind speed
Vibration
Biometric parameters

ECG
Electromyography
Frequency of breathing
Glucose and blood pressure
Heartbeat
Galvanic skin response
Body temperature

Table 1.
Variables actually measured by Scatol8.

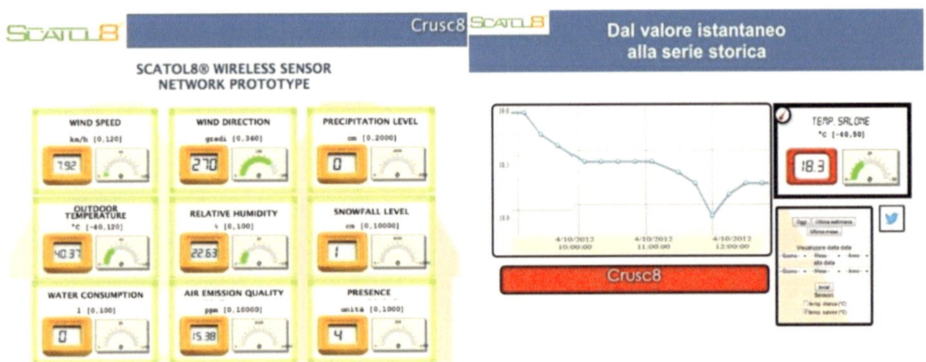

Figure 2.
Examples of Crusc8 and the temperature historical series.

1.2 Sustainable luxury products

Scatol8 srl products and services are manufactured according to the customer's specific requirements. Whether it is the planning and implementing of a remote sensing network, a training project, or a product, the answer to customer needs makes it a unique object. Everyone has the central idea to prove sustainability through materials and constructive methods. Therefore, materials are natural (various Italian woody species, natural fibers, and dyes) and the construction method crafts. From the union of these, unique and exclusive objects are created, whose sustainability is defined by scientific methods and processes. Features are tailored to the customer's needs, both in terms of sensors and actuators: A custom-made approach to sustainability.

The process shown in **Figure 3** highlights the steps that, from an idea, lead to the definition of product requirements after identifying and evaluating the conditions of economic, environmental, and social sustainability.

Being Scatol8 made of hardware, software, and containers, the definition of requisites analyzes, in a systemic way, the relations among these elements, to ascertain the feasibility of a product. The refinement work is carried out through the organized interaction of subjects with various training backgrounds, in a design thinking logic, which allows to explore various options for defining product requisites. Whenever a variation is suggested on one of the three elements to better respond to customer needs, the impact on other items is checked. This way of proceeding allows the definition of feasible product requirements. As far as the environmental sustainability test is concerned, life cycle assessment is the most commonly used methodology, while economic sustainability is examined with the business model

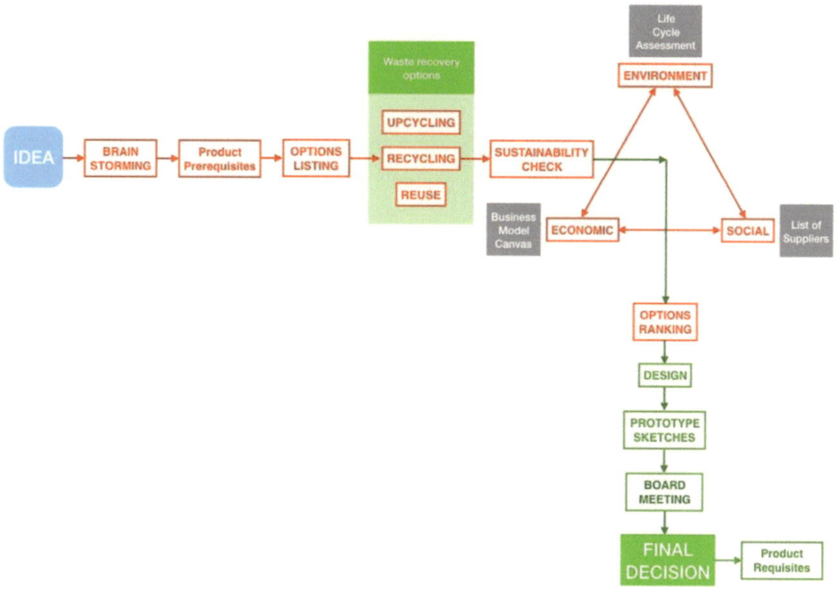

Figure 3.
The design planning from the initial idea to the definition of the product requisites.

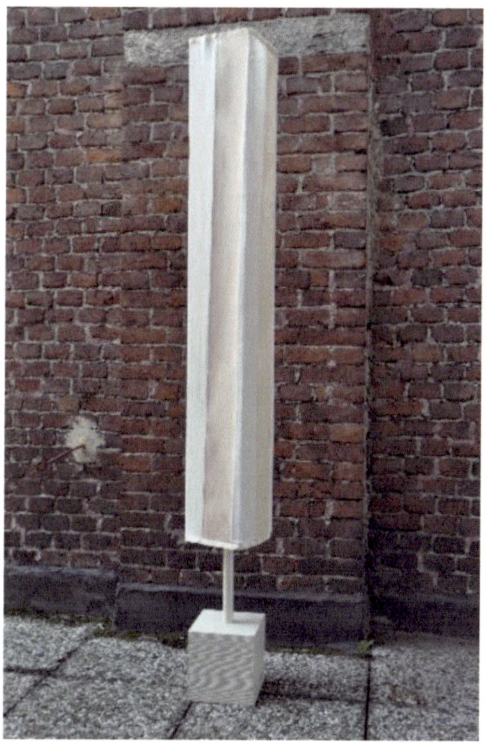

Figure 4.
The first Scatol8 lamp produced by Caneschi Ltd.: S8-LIL, Scatol8 Long Interactive Lamp.

canvas [1]. Social sustainability is sought through collaboration with companies located in Italy. Among the products that have been recently made, we present two data-knots, camouflaged by lamps.

Figure 5.
S8-SIL, Scatol8 Short Interactive Lamp.

Figure 6.
S8-vela.

Scatol8 revealed itself through prototypes, designed and produced in a green supply chain, in accordance with the strategic lines that guide our achievements. S8-LIL (**Figure 4**) and S8-SIL (**Figure 5**), Scatol8 Short and Long Interactive Lamp: lighting systems, in two sizes, to determine sequences bright, are dynamic interior design objects that contribute to well-being. Built of wood and natural fabrics, they interact with the environment, changing the appearance and promptly conquering the senses of the living space. Distributed in an apartment, S8-LIL and S8-SIL are a peripheral nodes of a Scatol8 network; the values of the environmental variables are conveyed to a router, published on the Internet, and displayed via smartphone and tablet. Lamps themselves (or any other object) can not only illuminate but also spread aromas, sounds, etc. It belongs to the same stream S8-Vela (**Figure 6**), consisting of a lamp, made of essence of chestnut, naturally dyed based on chestnut tannins, with medium-density fiberboard base, featuring three types of environmental monitoring sensors, in particular monitoring of temperature, humidity, and carbon monoxide parameters, also equipped with LED light actuators.

In the presentation of the Scatol8 system, detailed information on the hardware has been provided. The countless combinations and architectures of sensors, peripheral nodes, and coordinator node are not ends in themselves; indeed they are oriented to the collection and disclosure of information.

2. Wearable technologies

Wearable applications are possible thanks to the development of smart textiles, also known as electronic textiles, and thanks to the miniaturization of electronics that has allowed to integrate small circuits into the textile products [2]. This kind of textiles, introduced in the early 1990s, strongly influenced by military research and wearable technology in general [14], enables electronic components to be embedded in the fabric itself. The research behind these fabrics involves different topics like materials science, chemistry, design, and others. The multidisciplinary approach is also relevant because of the coexistence of design and usability of requirements. Another significant aspect in wearable technologies (also called wearable devices or wearables) is the optimization of power consumption in electronics that allowed to project smart applications whose batteries last enough for the specific usage. For example, the battery on a sport band must (at minimum) last the time of an average running session. New wireless technologies with low power consumption allow smart textile products to eventually communicate with other devices such as computer or mobile phones. In the case using these devices, the wearable application communicates on the Internet; we can say this product is also an IOT[2] application.

Wearable technologies can include many features. The smart textiles can be connected to sensors and react to external changes called stimuli. In response to these stimuli, the material can change some properties like its shape, color stiffness, etc. The wearable technologies can also include some actuators as some LEDs, displays, or speakers [3].

Virtually, in all the human cultures, people use garments not just for protecting themselves but also for distinguishing their sex, their social status, or their belonging to a specific subculture. Smart garments now extend their social function

[2] We can say that an object is IOT if it can be used without Internet connection, but within the net it extends its capabilities. For example, a server is not an IOT device because it is useless without a working net. On the other hand, a car can exist also without the use of the Internet to extend its capabilities, but the Internet would impact its nature or behavior, for example, giving real-time traffic information.

working as transmitters, sensors, actuator, or energy-harvesting entities. If these garments include a microcontroller, they can implement some logic like interacting with the cloud or with other devices [4].

Wearable technologies have been the subject of scientific studies since the mid-1990s [3]. The impulse, in terms of funding for research and product development, has been initially given by warfare applications. Immediately afterward, the potential of products in the biomedical sector and, subsequently, in sportswear area was understood. Lastly, with less resources, proposals have appeared in the fashion industry. The connections between these four sectors feed an increasingly rapid introduction of new products.

Presently, in the military sector, wearable technologies are identified as electronic technologies or computers that are incorporated into items of clothing and accessories which can comfortably be worn on the body. These wearable devices can perform many of the computing tasks as mobile phones and laptop computers; however, in some cases, wearable technology can outperform these handheld devices. Wearable technology tends to be more sophisticated than handheld technology available in the market today because it can provide sensory and scanning features not typically seen in mobile and laptop devices, such as biofeedback and tracking of physiological function [5].

Wearable technologies are now an important part in the process of developing military uniforms. New features can be monitoring the health of the soldiers as well as providing overall battlefield insights [6]. The military uses the data collected during engagement to improve the planning phase of future missions reducing causalities. One of the most important consequences using this kind of smart clothing is the reduction of injuries and consequentially the reduction of lifetime treatment borne by the whole country. The data collected from many soldiers and processed can help to dynamically change tactic as well as to improve the communication in the group [6].

Wearable military objects share many needs and requirements with some civilian applications. For example, in some sport or medical contexts, it would be useful to monitor some biometric parameter directly on the clothes. Devices like smart bands are already available on the market at cheap prices. New features are currently under development like monitoring athletes' oxygen saturation and heart rate and sending those data on a server where the coach and medics can analyze them [7]. Zhou et al. [8] explain a textile application where a matrix of sensors measures the muscle activities during the sport activities. In this way, it's possible not just to measure if there is a muscle contraction but also if the movement is done correctly. Another example in the sport context is a wearable system composed of a GPS connected with a breath sensor and an electrocardiogram to trace athletes' body performances [9]. Wearable sensors can be used not only for athletes; Mascia and others show an application where inertial sensors are used to evaluate people's movement, and in particular it is possible to check if a child can run properly [10]. Medicine can also take advantage from wearable technologies. Tuba and others explain how it is possible to detect vital signs with a sort of watch [11]. Dehydration can also be detected to evaluate the health status of elderly, diabetic, and sporting people in real time throughout the day [12]. Another interesting example of wearable medical application is a spine posture monitoring system based on inertial sensor. The data collected are elaborated with a mathematical and geometrical model; in this way it is possible to monitor rehabilitation sessions as well as to help patients to correct bad postures [13]. The sector of fashion has been the last one, in order of time and compared to the sectors cited before, to be involved in the IoT diffusion.

Lena Berlin, from the Swedish School of Textiles [14], did an in-depth analysis about smart textile in fashion. One first interesting point is that if medical and workwear application researches are mainly developed in public projects (e.g., EU projects) or universities, on the other hand sport applications and fashion textile projects are developed by universities and companies. After the literature analysis, the author said: An overall impression of the analysis and speculations is the strong belief that the potential of using smart textiles lies in application such as medical, workwear, and other technical applications rather than fashion. This can be true if we consider the potential of the research development, which is the main focus of scientific articles, but it is not so obvious if we consider the potential economic revenue in a huge market like the fashion one is. It's difficult to analyze that impact now because there are not still a lot of big companies working in fashion that have invested significantly in wearable. An example of a big company investigating in wearable fashion devices is Adidas that bought Textronics in 2008 [15]. There are also some technical barriers that are limiting the diffusion of smart textile applications. Sensors and radio can be quite expensive; for this reason many applications are just LEDs integrated in clothing. There is not a standard for interconnections between textiles and electronics, and it is difficult to make all the components robust enough. Because of the typical high costs and small production, we can say that wearable fashion products are not yet a mass product and mainly concern the luxury.

The wearable technology sector is, peculiarly, multidisciplinary. Considering innovations in tissues, research is a priority of materials engineering or bioengineering, with regard to hardware and software, electronic engineering, information technology, and telecommunications and with regard to shapes, architecture, design, etc. (see other references in the bibliography). If, then, it comes to assessing demand and defining production and commercialization strategies, we can see the involvement of product engineering, logistics, business economics, strategic marketing, etc.

The growth prospects of the wearable technology sector appear to be very promising, although there are substantial differences between sources. Recently, the "wearable technology market—global for 2022" report has predicted a growth rate of 15.51% between 2016 and 2022, moving from 15.74$ billion in 2015 to 51.60$ billion in 2022. However, it is a very wide sector, which includes also, but not only, textile products. Products that could have the highest growth rate are augmented reality glasses or virtual reality glasses. Hanuska and others show that the smart textile market for military uses is expected to significantly grow becoming a 500 M$ market in 2018 because this is now a common need shared among almost all the countries in the world.

The EU has supported research in the wearable sector since 2002, allocating 3.984 million with the FP6 program. Subsequently founding continued with FP7 program. The main projects are [16]:

1. "My Heart," 16 million/allocated from 2003 to 2009

2. "Dephotex," funding of 3.131.482 Euros

3. "Stella Project," funding of 7 million Euros

The biomedical sector was the most funded one by the EU. Despite an extensive research effort in several projects for over 10 years, there are only few smart textile clothing products on the market, and the volume of business, if declared, seems to be modest in the context of fashion and clothing. However, there are some newly established companies focused on the development and commercialization of smart textile clothing. An interesting aspect in these efforts to commercialize smart textiles is the interdisciplinary collaboration between companies in fashion and electronics, respectively. Besides pure fashion companies, there are some companies established that sell know-how on how to integrate electronics into textiles and clothing. A number of EU projects in smart textiles have been supported over the last decades. Most of the supported projects are within the health monitoring area. Another type of projects at the EU level is developed enabling technologies for smart textiles, for example, stretchable electronics, integration of electronics in textiles, technologies that are necessary for the development of smart textile applications, health monitoring for medical assistance, health monitoring integrated in work wear, projects developing enabling technologies, fashion, and clothing companies.

Consultancy partners specialized in textile and electronics. The combination of textiles and electronics in smart textiles has opened up for a new type of consultancies who are specialized in the combination of textiles and electronics. These consultancies concentrate their business in supporting other companies in their manufacturing of smart textiles and clothing rather than manufacturing and selling their own collections. However, some of them combine their consultancy with the manufacturing and marketing of their own technologies or materials [14].

2.1 Wearable technologies and fashion

The Swedish school of textile has thoroughly investigated the field of wearable technologies. The report on smart textiles and wearable technologies [6] presents an overview as a basis of further discussion of how smart textiles could be introduced in fashion. As shown there are already some commercial initiatives around Europe who specifically target fashion. What is also obvious is that there has been an extensive research activity both at the European and national levels in the area of smart textiles and clothing. The total funding of the presented EU projects, for example, is around 70 million Euros, which could be seen as a high financial contribution. It should though be noted that the money is shared between researchers in different areas such as textiles, electronics, wireless technology, battery research, and system engineering. These research efforts are therefore not only a concern in smart textiles since the results also contribute to developments in other areas. Despite a rather extensive research effort, the industrial and commercial activities are still in its infancy.[3]

The report "a roadmap on smart textile" [17] focuses on the potential of the smart textile; it divides the market into three areas: healthcare, workwear, and sports. The fashion industry is not studied and neither considered as a field of interest for the future. In another market survey, carried out by Ohmatex [18], fashion is considered as one of the areas but assessed as irrelevant initiatives, because of the size of the impact. The analysis reveals the barriers between the research and the commercial outlets which have been overtaken by several emerging designers, among which, for the originality of the proposals, Cutecircuit [19, 20], Pankaj and Nidhi [21], Anouk Wipprecht [22], Becca McCharen [23], Pauline van Dongen [24], and Akll Giysiler [25]. Some products result in an effective connection between healthcare and workwear, mediated by innovative stylistic elements: Clothing+, HVDING [26], MOON Berlin [27], Myontec, No-contact [28], Philips Lightning Lumialive [29], Stealth Wear anti-drone garments [30], Textronics [31], Utope [32], and WARMx [33].

The difference between implementations for healthcare and workwear, and implementations in the fashion field, is that in the first case, they are oriented to health monitoring or facilitating the wearer's communication, whereas in fashion applications, they are more oriented to a visual or a tactile feedback.

Looking at the latest high-fashion brand collections, which include smart clothing or accessories, we can find encouraging signals about the business perspectives. Ralph Lauren, Opening Ceremony, Rebecca Minkoff [34, 35], Karl Lagerfeld, and Hussein Chalayan [36] are just some of the great brands that have integrated various types of wearable technologies into their products. Chanel [37] launched a line of LED-powered handbags in the Spring-Summer 2017 collection, and it appears to be

[3] Lena Berglin, op. cit.

Figure 7.
One "ready to wear" kimono.

determined to integrate new technologies into its brand. Chanel has been trying to sort out its relationship with technology, looking for the ways in which technology tucked into clothes can improve our lives or at least make fashion become more interesting.

The contemporary presence of emerging designers and "approaches to wearable technologies," made by big brand, Andrew Bolton, the Costume Institute's chief curator, states that "Technology is eroding the difference between haute couture and ready-to-wear" [38].

The most common market entry barriers are current technology limits and the lack of standardization. Encumbrance, energy consumption, interconnection between textile materials and electronics, lack of standardization and quality systems, and training are examples of technical barriers. Low production rates and costs are barriers which are additional aspects related to the constraints imposed by safety and health. These factors affect both emerging designers, oriented to design and produce exclusive outfit and prototypes, and high fashion, oriented to realize one-of-a-kind outfit (**Figure 7**). It is likely that collaborations between these worlds, distant from economic capacity but convergent on technologies and the scale of production, can be considered. The adoption of wearable technologies by the industry, as well as it is for the society, is just at its first steps. Mesut [39] lists some of the possible applications: ease the life for the people with impairments; enable companies to interact with theater business people easier, to conduct market research more effectively and to apply sales and service strategies more efficiently; enable policemen, firemen, and military members to provide public and personal safety; enhance the virtual reality in games; and enable the doctors to monitor health indicators for the people continuously. These examples demonstrate how wearable technologies will make life easier and safer and how they can help to improve the entertainment market with new features.

3. The Scatol8 srl in the computational clothing

Scatol8 srl has a branch in the textile industry, called *Indigo Laboratories*. The emphasis on laboratory activity refers to the performance of the garment creation; in particular they are carefully designed to achieve both esthetic and functional performances. In the textile field, new technologies allow to extend the functions of garment. Indigo Laboratories, with its products, operates on the integration between electronics and textiles. The business is the *creation of prototypes* that incorporate the Scatol8 guidelines (accessibility, modularity, eco-compatibility, dissemination of knowledge) and their application to the production of *garments integrated with hardware and software*, components belonging to the *wearable technologies*, or, more specifically, *computational clothing*. The commitment in this area is functional to the achievement of Scatol8's goal, which is to ultimately disseminate the culture of sustainability, through modular and environmentally friendly products and projects, which promote relationship awareness between the individual and the environment where he lives. Therefore, our proposal combines a scientific application—related to the monitoring of environmental quality parameters—with fashion.

"Computational clothing" is a concept which emphasizes efforts to integrate computers and clothing. The manufactory aims to make garments which have the ability to process, store, retrieve, and send information. This capability will allow clothing and accessories to work as a stand-alone computer, to "react" according to environmental conditions and/or to connect to the Internet or other networks. Also, computational clothing will be able to change its appearance; it will allow users to access on specific apps "loaded" on modern smart devices (smartphones, tablets, etc.) [40].

In the field of textiles and clothing, *Scatol8 srl works in a network with some close relatives*:

- With the *QUMAP Laboratory* (QUMAP is the Italian acronym for quality of goods and product reliability), located at the PIN—University of Citta di Prato.

- With the *Phytolab-DiSIA Laboratory* of the University of Florence, to study and applicate new plant fibers (e.g., nettle and hemp) and natural dyes obtained from waste of the agro-industry such as oenological, fruit and vegetable, olive oil sector, and/or by officinal dyeing species.

- With two Italian SME that produce eco-leather from vegetable source, that is, the scraps of vine-wineries and olive tree pruning activities.

3.1 Interactive denim kimonos

The capsule collection we made, interactive denim kimonos (**Figure 8**), is made up of five short kimono models (Hanten), which respect the traditional proportions of this type of garment, winking to a more European fit, so it can be easily worn all days, in casual circumstances. The choice of a garment that belongs to the millennial Japanese tradition, but realized with the renowned Italian craftsmanship, has been carried out with the specific aim of demonstrating how the integration with wearable technologies is more than ever flexible and can be adapted and modulated according to the context, even if this means merging two apparently distant worlds.

Kimonos are packed in an environmentally friendly way: each garment is made from three vintage high-quality jeans (100% cotton) that, dismantled and reassembled, create a new item of clothing with completely different use of the starting materials, even though it has their distinctive patterns in a patch of blue or black tones.

We used special natural fabric ribbon, dyed with natural dyes for the integration of LEDs and patches in eco-leather, it has harmoniously conversed with the jeans patchwork, and it allowed us to achieve a unique and original result. Natural dyes, obtained from aqueous extraction or by green technologies, may be yellow obtainable from *Carthamus* or *Curcuma*; yellow-orange from golden onion and red onion; red from species rubia red beet, and myrtle or from scraps of the wine sector such as vinegar and grapevine; brown obtainable from chestnut extract; blue from spirulina algae; or by extraction from the Guado species.

Alternatives of hardware, software, and fabrics have been routinely verified in their reciprocal relations, to reach the present proposal by a network of Italian companies that worked together for the execution of the capsule collection.

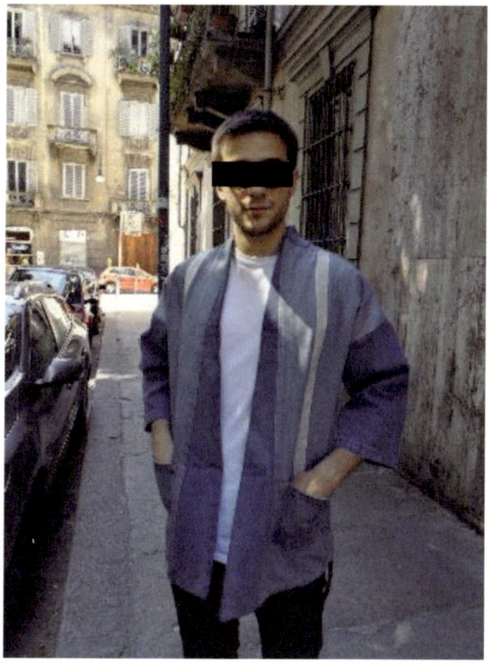

Figure 8.
One example of interactive kimono.

3.2 Hardware and software components

Kimonos are equipped with a Scatol8 system, made up of hw and sw. The hardware consists of the following elements:

- Microcontroller Arduino-Arduino Micro
- Bluetooth module DSD TECH HM-10 Bluetooth 4.0 BLE IBeacon UART with 4PIN Basic card for ONU R3 of Arduino 2560 mega nano-sensors
- NooElec 1m actuators 60-Pixel Addressable 24-Bit RGB LED Strip, 5V, Waterproof IP68, WS2812B (WS2811), 4-Pin JST-SM

The *software* runs the Scatol8 system:

- Allows dialog between microcontroller, sensors, and actuators
- Realizes the smartphone app in Android and iOS
- Manages communication, via Bluetooth, to the smartphone, allowing colors changing schemes and sequences of the LEDs
- Manages communication between Scatol8 and the server
- Draws and makes the dashboard Crusc8 working
- Processes data to draw environmental quality maps

Table 2 shows the programming languages used for carrying out the activities.

Project part	Adopted language
Microcontroller	C++
App mobile	C#, sql
Database	Sql
Backend	HTML, PHP, JavaScript, SQL

Table 2.
The adopted programming languages.

Figure 9.
A running kimono. It detects one environmental variable and change the LED's light proportionally to the acquired value.

Features can be set based on the user's needs both with sensors and actuators. The prototypes presented change their appearance depending on the sound intensity or the temperature or the brightness; dozens of sensors are available, and most of them are currently being tested to customize kimonos. The LEDs, driven by the open-source microcontrollers, take on a bright intensity and different colors to reflect the trend of the variables and to adapt to the color of the fabric (**Figure 9**). In addition to the LEDs, there are other actuators used in computational clothing: for example, it is possible to use heating systems that intervene when the temperature drops below a certain threshold, interaction systems with smartphones, for the collection and transmission of data, buzzers, etc. Also, through a smartphone application (**Figure 10**), you can *set the color and LED sequences* as you liked, depending on the style of your cloth, circumstance, or mood. In this way, the system continues to collect data through the sensors, but the chromatic effect is not determined by the intensity of the monitored variables. Kimono holders can log and make available the date, time, place, and intensity of the variable measured by their clothing (**Figure 11**). This allows the drawing of environmental thematic maps.

3.3 Environmental review

Kimonos have sprung out from the design planning of **Figure 3**. Alternatives of hardware, software, and fabrics have been routinely verified in their reciprocal

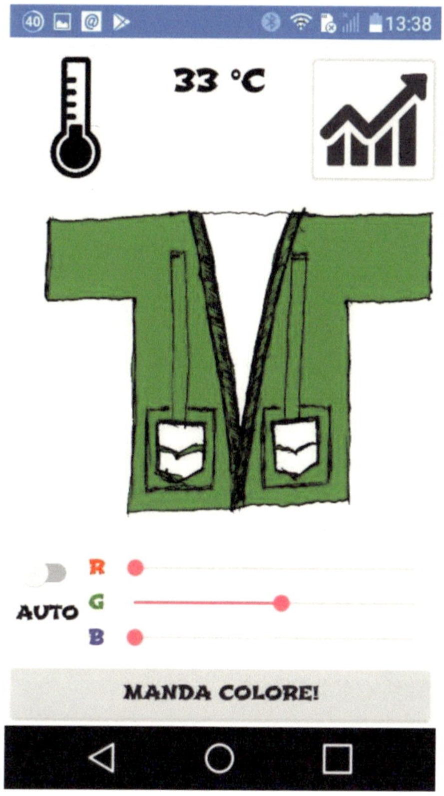

Figure 10.
The mail page of the application. The user can set the color of the kimono and read the detected environmental variable.

Figure 11.
The smartphone app forwards the collected data to a server. In this way it's possible to check the values everywhere using a browser, not just if you are near the kimono.

relations, to reach the present proposal. **Figure 12** describes the workflow of an interactive kimono. If we consider the process for the production of kimono from a gate-to-gate perspective, the input factors are:

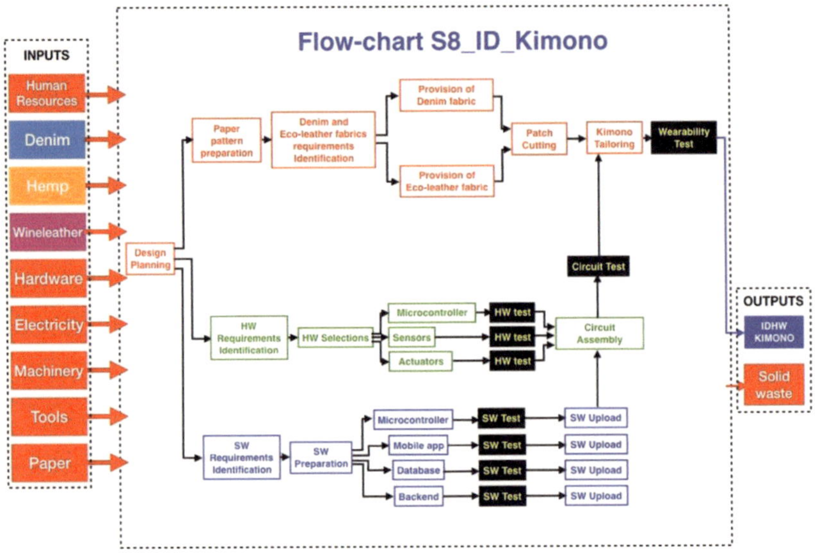

Figure 12.
The workflow of an interactive kimono.

- Fabrics (denim, hemp, eco-leather)
- Various materials (sewing thread of various diameters, model paper, writing instruments, tailor-made meter)
- Equipment (sewing machines, computers, microcontrollers, sensors, LEDs)
- Tools (tailor's scissors, needles)
- Various software
- Electric energy
- Human resources of multiple professionalism

The operations carried out are exclusively of a physical nature (drawing, cutting, and sewing), and the wastes (wastes of fabric) are reused in other artifacts.

Noting the low environmental impact of the operations carried out in the production of kimono, the improvement of environmental performance has been sought through a critical analysis and evaluation of kimono components, considering the type of raw materials and components and the susceptibility to substitution, in relation to the available environmental information. **Table 3** describes the situation.

Following the implementation of this decision-making scheme, in the framework of an environmental review, attention was focused on the type of fabric and on the identification of an alternative to cotton use.

With regard to the opportunity of denim replacement, this has emerged with reference to the environmental impacts that take place upstream within the Indigo Laboratories production process.

From a broader point of view, from cradle to grave, the life cycle assessment (LCA) would take into account the environmental impacts of denim.

Parts of kimono	Need for replacement	Susceptibility to substitution
Denim (cotton)	Appropriate	Possible
Ribbon (natural fibers dyed with natural dyes)	Not necessary	—
Patch of eco-leather	Not necessary	—
Electronics	To be verified	Not possible in the short term
Batteries	To be verified	Partial

Table 3.
Parts of a kimono and opportunities for ad environmental improvement.

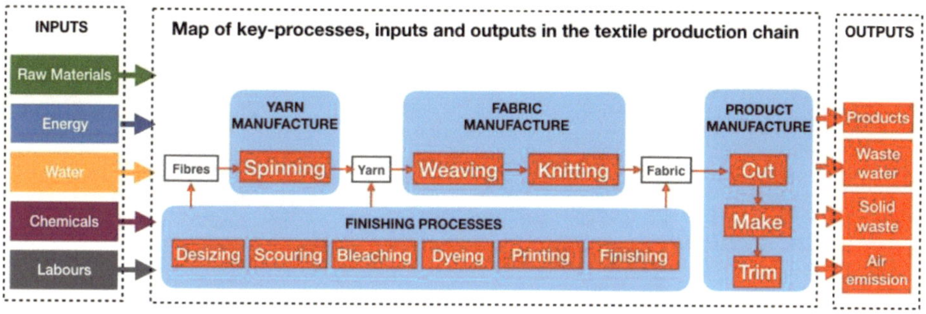

Figure 13.
The map of key processes.

Figure 13 shows the map of key processes in the textile production chain, no matter the type of fiber, being it synthetic, artificial, or natural.

Fibers, yarns, fabrics, and garments are the elements that mark the evolution stages. Each of them is a product, i.e., the output of transformation activities that generate a burden of an environmental impact, in terms of waste water, solid waste, and air emission.

If you want to evaluate the complete environmental impact, applying the LCA methodology, upstream the manufacturing process, there is the stage of fiber production; downstream there are the stages of distribution, use, and disposal [42].

Figure 14 drops the general scheme to the case of the production of cotton clothing. Each phase can be considered in itself a from-gate-to-gate form, or the whole cycle of transformations can be evaluated, in a from-cradle-to-grave form. *Scott Camp*, *Gordon Clark*, *Laura Duane*, and *Aaron Haight* have studied the life cycle of jeans from cradle to grave as a collection of seven independent systems. These systems in their respective order are (1) cotton production; (2) fabric production; (3) garment manufacturing; (4) transportation and distribution; (5) consumer use; (6) recycling, which then goes back to step (2); or (7) waste stream in a landfill [41]. Each phase is linked to the subsequent by a transportation activity. If we consider that cotton cultivation is restricted to subtropical areas and that garment manufacturing involves companies located in various countries, from Far East to South and Central America and the USA, it's easy to have an idea about the environmental burdens of transports.

Figure 15 deepens the disposal phase and includes recycling processes. After the use phase, discarded garments can be incinerated (with the generation of energy) or dumped into a landfill. But they can be recycled, after collection and sorting. Depending on their conditions, they can be reused as secondhand products; their

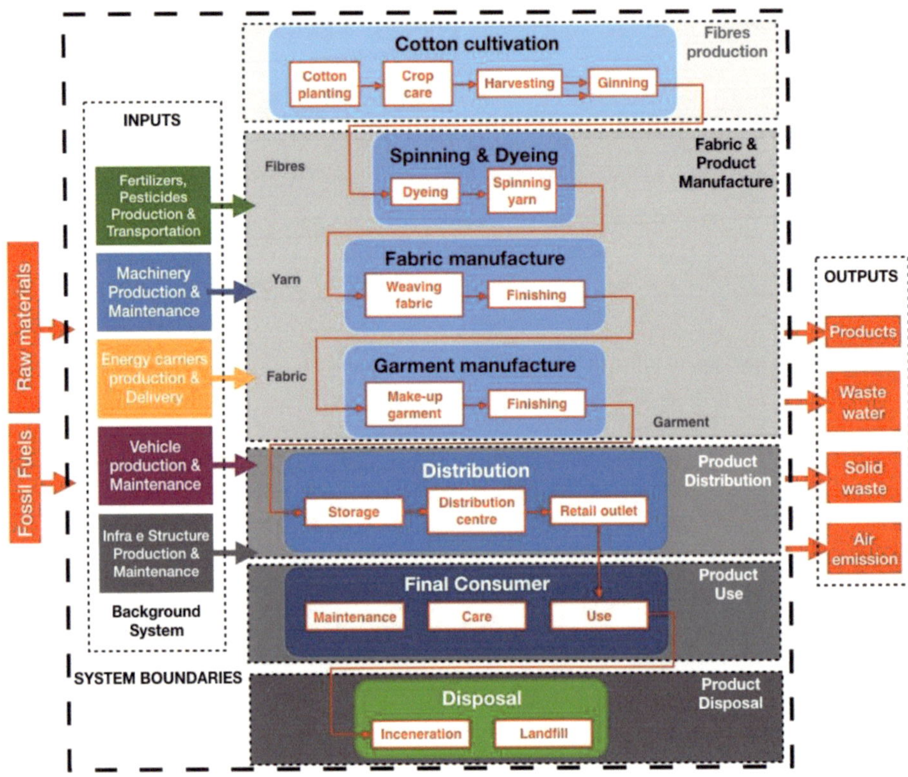

Figure 14.
Life cycle stages of a cotton clothing.

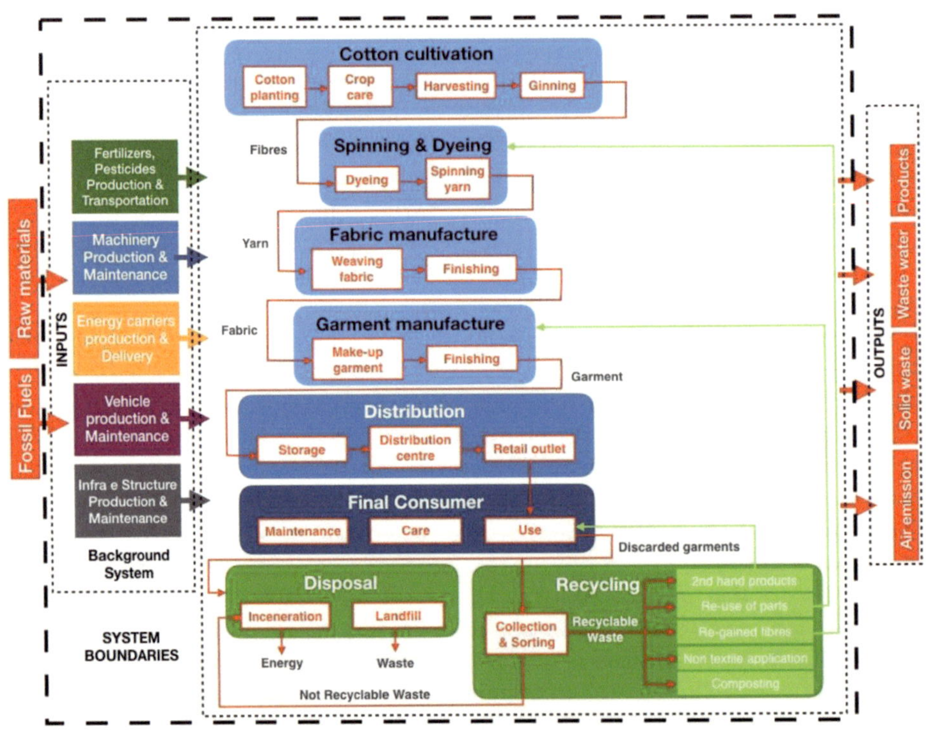

Figure 15.
Life cycle stages of a cotton clothing, with a focus on end-of-life options.

fibers can be recycled in the production of a new fabric; parts of them can be used for the production of new garments; and, in this case, we use the term *upcycling*.[4] The valorization of discarded garments is made up of the circular economy and extends the life cycle of a product, promoting material and energy recovery.

With regard to denim (cotton), the recovery of a material results in an environmental gain as disposal has been avoided; the same can be said for the eco-leather fabrics, being produced with waste of agriculture.

Numerous bibliographic references on the application of LCA to cotton fiber, denim fabrication, jeans packaging, use, and disposal emphasize critical aspects in various phases, especially in relation to water use and use of pesticides.[5]

Cotton is the world's most pesticide- and chemical-intensive crop. Of all the insecticides sprayed worldwide, cotton spraying accounts for roughly of them, also consuming 10% of the world's pesticides on an annual basis with an estimated cost of the pesticides totaling $2.6 billion [43].

These factors have been highlighted and discussed in the framework of a review by the Indigo Laboratories team with the aim of identifying improvement margins and assessing their feasibility.

The research of an alternative fiber was based on scientific contributions that evaluated, with LCA methodology, the environmental impact over the entire life cycle of other fibers.

Considering the whole textile chain, from spinning to finishing, it cannot be ignored that the use of chemicals may have carcinogenic and neurological effects, may cause allergies, and may affect fertility. During these processes, large amounts of water and energy are used, and, in general, non-biodegradable wastes are produced [55].

What are the alternatives to cotton? Emily Kenny-Troughton [46] examines all the available options to avoid or limit the environmental and social impact of the denim production process (**Figure 16**). These alternatives include, but are not limited to, recycled cotton, in-conversion cotton, organic cotton, organic flax, bamboo fiber, BCI cotton, hemp and nettle fibers, TENCEL TM as apex Modal, and recycled polyester. Conventional hemp is very different from cotton; as the plant requires little to no pesticides, fungicides, or herbicides and as it grows so

[4] With the term upcycling, the transformation of a waste into a new fashion object using creativity is indicated. Coined for the first time in 1994 by journalist Reiner Pilz and officially cleared in 1997 in the same book by Gunter Pauli, the concept of upcycling is well defined and largely distinct from the most consolidated recycle term, which describes an industrial process of transformation of waste. The "end of life" of products in the fashion system Rome, October 2013, edited by Clemente Tartaglione and Sara Corradini with research contributions of Gianmarco Guazzo, Mauro Di Giacomo.

[5] A large part of the sustainable solutions currently offered focus on the fibers used to create the garments and the ethical and environmental impact that they have. Aside from the petroleum-based obvious bad guys like polyester, cotton is one of the least sustainable fibers currently in use by the clothing industry. Issues with the fiber range from enormous water usage to the controversy of GMO crops and from exploitation of farmers to the widespread use of harmful chemicals. Around 20 million tonnes (USA) (18.14 billion kg) of cotton are grown every year (wwf.panda.org, 2016), and these fibers are present in over 50% of all clothing and other textiles (cottoninc.com, 2016). An example of how inefficient growing cotton is takes around 1514 liters of water for a simple cotton t-shirt (including all processes) and around a staggering 6814 liters of water (Tree- Hugger.com, 2016) to fully process a pair of jeans. The amount of cotton needed for one t-shirt and one pair of jeans is 1 kg (wwf.panda.org, 2016), and just over 1 billion pairs of jeans are sold annually on a global scale (statisticbrain.com, 2016), which gives a rough total of around 1 billion kilos of cotton being affected by the denim industry a year. This shows that steps taken to reduce denim's impact can have large-scale consequences. Research Report, Emily Kenny-Troughton, 500668689, International Fashion & Management, Ligia Hera & Jacqui Haker, page 5, 13/06/2016.

Figure 16.
One kimono made of hemp.

fast, it leaves all other weeds in the shade. In order to make threads from hemp, it is necessary to use the best fibers that are found in the stalk of the plant. This process does not require any chemicals at all, only using the enzymes naturally found in the plant itself meaning that it does little to no harm to the environment, workers, or end users of the product [45]. Lea Turunen states that comparison between hemp and cotton is difficult, due to lack of comparable data, but for the crop production stage, hemp performs clearly better than cotton with respect to pesticide use and water use. To improve hemp performance, she recommends to concentrate on fiber processing and yarn production stages, where hemp requires more energy [47].

There are studies which provide rankings of textile fibers according to their environmental impacts, as shown in **Table 4** and **Figure 17** [44, 45].

Hemp conventionally grown performs very well. In addition, its evaluation would be even better in our reality, as hemp has been cultivated for centuries in

Ranking	Fibers
Class A	Recycled cotton, recycled nylon, recycled polyester, *organic hemp*, organic flax (linen)
Class B	TENCEL, organic cotton, in-conversion cotton
Class C	*Conventional hemp*, ramie, PLA, conventional flax (linen)
Class D	Virgin polyester, polyacrylic, Lenzing Modal
Class E	Conventional cotton, virgin nylon, rayon (cuprammonium), bamboo viscose, wool, generic viscose
Unclassified	Silk, organic wool, leather, elastane, acetate, cashmere, alpaca

Table 4.
Environmental impact of textile fibers.

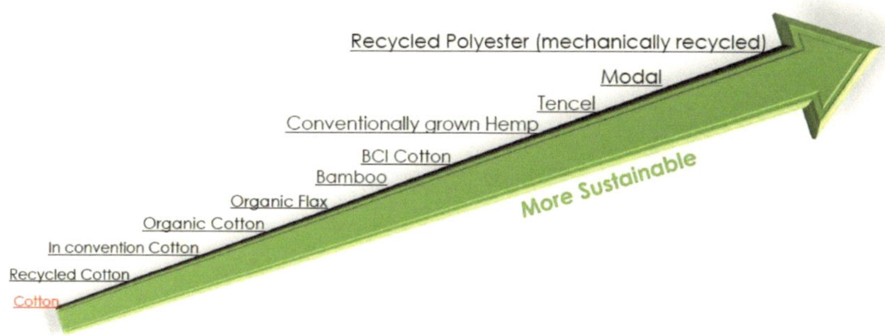

Figure 17.
Degree of sustainability of various fibers compared to cotton.

Piedmont. The fabric supply market is very close to our company, minimizing economic and environmental impacts.

Figure 18 shows the production of hemp fibers. Spinning and dyeing, fabric manufacturing being the same as cotton, and differences in terms of environmental impacts can be found at the agricultural production, straw processing, and transport stages. The best alternative is raw hemp or dyed with natural pigments.[6]

In literature, many studies highlight the benefits of hemp, also due to its use in many economic sectors, leading to the integral use of its parts [47–51].

It is a fiber that can be used for the production of a variety of commercial items such as textiles, clothing, thermal insulation, paint, paper, biofuel, biodegradable plastics, food, and animal feed [52, 53].

Ribbons and patch of eco-leather—The ribbons and the patch in eco-leather are made of natural fibers and dyed with natural pigments; they do not need to be replaced.

Electronics—Electronic devices, consisting of microcontrollers, sensors, and LEDs, evolve rapidly: it is likely to presume that the market will produce even more miniaturized products, available, so constructed with less material, as a concrete example of dematerialization strategies.

To get an environmental impact assessment of electronic components, we started an LCA study of the electronic devices we use. As far as LEDs are concerned, there are studies that unequivocally determine environmental benefits of LEDs compared to other forms of lighting (which would not be usable in

[6] Conventional hemp is very different from cotton; as the plant requires little to no pesticides, fungicides, or herbicides and as it grows so fast, it leaves all other weeds in the shade.

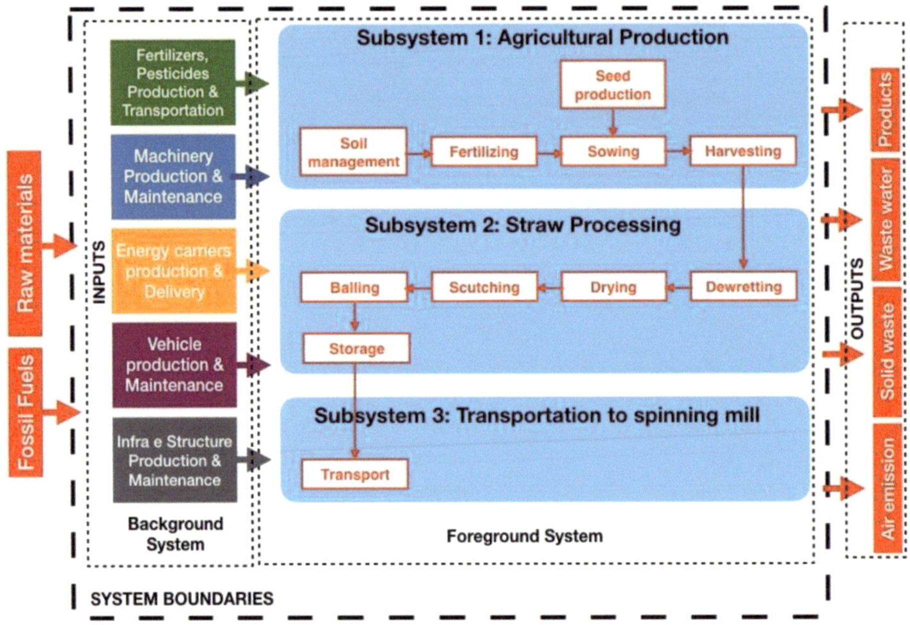

Figure 18.
Production of hemp fibers.

garments); the literature does not provide estimates for microprocessors and sensors that we used.

Batteries—The batteries currently used have a weight of about 250 g and a variable lifetime depending on the type of usage but can also reach 3 h of autonomy before being recharged.

The ability to integrate into the garment photovoltaic cells would reduce the size and weight of the batteries but would require an LCA verification of the cells themselves. It is something on which we are currently working with a special prototype.

As a final decision, at the end of the review on environmental performance of kimono, it has been decided to plan the manufacture of a line of hemp made of kimono which is significant not only for the environmental point of view but also for the recovery of cultural values associated with the cultivation and use of hemp.

4. Conclusions

Scatol8 for Sustainability is an entrepreneurial initiative that raises the theme of sustainability by incorporating it into luxury goods. Buying a Scatol8 product means belonging to an elite who appreciates handicraft products, an expression of Made in Italy, and acting to be aware of the environment in which one moves. Then, culture becomes a matter of luxury. In a magmatic market of consumer goods, it is not the price to decree the exclusive connotation. The price limits the boundary between who has and who does not have. Luxury, in the Scatol8 vision, delimits the boundary between who is and who is not.

The concept of Scatol8 has been presented through various luxury products (**Figure 19**) that testify the creativity of our innovative start-up. Multidisciplinary know-how, creativity, innovation, short lead time, and outcomes that exceed expectations are the critical success factors we are working on to improve ourselves continually.

Figure 19.
The capsule collection S8-ID-kimono, presented during the exhibition "Y Kimono Now," in Caraglio (CN), Piedmont, Italy.

The *S8_ID_KIMONO* has been analyzed in detail; it fulfills the requirements of portability, modularity, and awareness-raising through continuous monitoring of environmental conditions. Hardware and software have been optimized to reduce size and energy consumption. *Indigo Laboratories' proposal incorporates functional and fashion aspects. The functional side is taken from the various electronic devices and consists in monitoring environmental variables; the fashion part is represented by the colors, their variation, and intensity, which transform the garment*. The transmission and processing of data add a scientific aspect. Regarding the monitoring of parameters related to the quality of life, the introduction of new sensors on air quality and the biomedical field will broaden the capability of the garments and, consequently, the business opportunities.

Indigo Laboratories intends to expand the number of corporate network members, to increase the effectiveness of joint ventures. Starting with the design of garments, developed together with fashion institutes, thanks to a diversification of production cycles, Scatol8 will be able to develop products whose degree of customization will increase. In view of continuous improvement, we plan to introduce patches of natural fabrics, dyed with natural dyes, to enhance cultivation and environmentally virtuous practices, carried out in Italy.

In addition, a current work in progress is the expected adoption of design for environment techniques to develop garments that, from design, propose to integrate hardware not to simply add it. The transition from application to electronic integration is important, in our opinion, to give birth to really new cloth in shapes and style.

In addition to kimono, the electronic system miniaturization and the versatility of the LEDs allow Scatol8 to be integrated in other garments, suitable for every occasion; for example, the cotton t-shirt collections, which supported Scatol8's message from 2013 onward, made with strictly certified sustainability fabrics [54], will be renewed and revolutionized by the interaction made possible by the patented device.

Cost reduction that could be achieved by standardizing the models would lead to the spread of garments that, from the point of view of environmental quality research, would potentially allow thousands of surveys, georeferenced, useful to undertake health policies and to evaluate their effectiveness. Monitoring of environmental conditions and biomedical parameters will provide guidance to improve the quality of life by enabling, in emergencies, the sending of warning signals or the pilotage of systems designed to improve conditions.

Author details

Riccardo Beltramo[1*], Jacopo Beltramo[2], Paolo Cantore[1] and Annalisa Romani[3]

1 Department of Management, Area of Commodity Science—Lo Scatol8 per la Sostenibilità srl, University of Turin, Turin, Italy

2 Lo Scatol8 per la Sostenibilità srl, Indigo Laboratories, Turin, Italy

3 QUMAP and Phytolab-DiSIA Laboratory—Lo Scatol8 per la Sostenibilità srl, University of Florence, Florence, Italy

*Address all correspondence to: riccardo.beltramo@unito.it

References

[1] Beople. The business model Canvas. 2017. Available from: http://www.businessmodelcanvas.it/

[2] Berglin L. Smart Textiles and Wearable Technology—A Study of Smart Textile in Fashion and Clothing. Swedish School of Textiles, University of Borås; 2013. p. 7

[3] Henock HD. Literature over view of smart textiles [Thesis for the Master of Textile Technology]. Boras, Sweden; 2010

[4] Foroughi J, Mitew TE, Ogunbona PO, Raad R, Safei F. Smart fabrics and networked clothing: Recent development in CNT-base fibers and their continual refinement. IEEE Consumer Electronics Magazine. 2016;**5**(4):105-111

[5] National Institute of Advanced Studies. International Strategic and Security Studies Program (ISSSP), Wearable Military Technologies, ISSSP Reflections No. 30, Bangalore, India; 2015

[6] Hanuska A, Chandramohan B, Bellamy L, Burke P, Ramanathan R, Balakrishan V. Smart Clothing Market Analysis. Berkley: Berkley University of California; 2016

[7] Fua Y, Liub J. System design for wearable blood oxygen saturation and pulse measurement device. Procedia Manufacturing. 2015;**3**:1187-1194

[8] Zhou B, Sundholm M, Cheng J, Cruz H, Lukowicz P. Measuring muscle activities during gym exercises with textile pressure mapping sensors. Pervasive and Mobile Computing. Sep 2016. DOI: 10.1016/j.pmcj.2016.08.015

[9] Li RT, Kling SR, Salata MJ, Cupp SA, Sheehan J, Voos JE. Wearable performance devices in sports medicine. Sports Health. 2016;**8**(1):74-78

[10] Mascia I, Vannozzia G, Bergaminia E, Pescea C, Getchellb N, Cappozzoa A. Assessing locomotor skills development in childhood using wearable inertial sensor devices: The running paradigm. Gait and Posture. 2013

[11] Yilmaz T et al. Detecting vital signs with wearable wireless sensors. Sensors (Basel, Switzerland). 2010;**10**(12):10837-10862. DOI: 10.3390/s101210837

[12] Zak J, Solovei D, Sedlacek J, Hubalek J. Development of wearable medical system for dehydration detection. 2012 4th Electronic System-Integration Technology Conference, ESTC. 2012. pp. 1-6. DOI: 10.1109/ESTC.2012.6542216

[13] Voinea G-D, Butnariu S, Mogan G. Measurement and geometric modelling of human spine posture for medical rehabilitation purposes using a wearable monitoring system based on inertial sensors. Sensors. 2017;**17**(1):1-19. Article No: 3

[14] Berglin L. Smart Textiles and Wearable Technology. The Swedish School of Textiles, University of Borås; 2013

[15] Rutherford JJ. Wearable technology. IEEE Engineering in Medicine and Biology Magazine. 2010;**3**:19-24

[16] Community Research and Development Information Service—Project and Results Site. 2017. Available from: http://cordis.europa.eu/projects/home

[17] Anne S, van Langenhove L, Guermonprez P, Deguillemont D. A roadmap on smart textiles. Textile Progress. 2010;**42**(2):99-180

[18] Ohmatex project. 2006. Available from: http://www.ofseth.org

[19] Wearable technology. Vision and history. 2017. Available from: https://cutecircuit.com/wearable-technology/

[20] Twitter Dress: Social...fashion. 2017. Available from: http://www.bigodino.it/tecnologia/twitter-dress.html

[21] Pankaj, Nidhi. 2017. Available from: http://www.pankajnidhi.com/ss12-Geometrica.html

[22] Synapse dress, l'abito smart che veste il modo in cui ti senti. 2014. Available from: http://www.medaarch.com/2014/09/synapse-dress-labito-intelligente-che-veste-il-modo-incui-ti-senti/

[23] Codame Fashion+Tech. Behance. 2017. Available from: https://www.behance.net/gallery/16246789/Codame-FashionTech

[24] Werable Technology. From Fitbit to Solar Cells. 2014. Available from: http://www.dressed.so/blog/2014/08/08/wearable-technology-from-fitbit-to-solar-cells/

[25] Giyilebilir teknomoda: Akilli Giysiler. 2014. Available from: https://www.youtube.com/watch?v=ordT391jXeY

[26] Hovding. 2017. Available from: https://hovding.com/

[27] Moon Berlin. 2017. Available from: http://www.moonberlin.com/

[28] No-contact. 2017. Available from:http://www.no-contact.com/

[29] Philips showcases production-ready Lumalive textile garments. 2006. Available from: https://phys.org/news/2006-08-philips-showcases-production-ready-lumalive-textile.html

[30] Ahproject. 2017. Available from: https://ahprojects.com/projects/stealth-wear/

[31] Textronics. 2017. Available from: http://www.textronicsinc.com/

[32] Utope. 2017. Available from: http://www.utope.eu/sporty-supaheroe jacket01.html

[33] Smart textiles—product development around textile current circuits. 2017. Available from: http://www.warmx.de/index.php/industry-and-research.html

[34] Rebecca Minkoff. 2017. Available from: https://www.rebeccaminkoff.com/

[35] The rise of wearable technology. 2015. Available from: https://dispatchist.com/the-rise-of-wearable-technology-interlaced/

[36] Paris Fashion Week, sfilano occhiali e cinte smart di intel. 2016. Available from: http://www.corrierecomunicazioni.it/it-world/43735paris-fashion-week-sfilano-occhiali-ecinte-smart-di-intel.htm

[37] Weeables on the Runway. 2016. Available from: https://www.wearable-technologies.com/2016/11/wearables-on-the-runway/

[38] Brooklin's Wearable Revolution. 2016. Available from: https://www.nytimes.com/2016/05/01/fashion/brooklyn-wearables-revolution.html?

[39] Mesut C. Wearable technologies and its future applications. In: Proceedings of ISER. Science Plus International Conference; Bangkok, Thailand. 2015. ISBN: 978-93-84209-99-5

[40] Barfield W, Mann S, Baird K, Gemperle F, Kasabach C, Stivoric J, et al. Computational Clothing and Accessories. 2001

[41] Camp S, Clark G, Duane L, Haight A. Life Cycle Analysis and Sustainability Report—Levi Strauss & Co.—ENVS 195: Sustainability Science, Dr. Saleem Ali. 2010

[42] Tartaglione C, Corradini S, Guazzo G, Di Giacomo M. The "End of Life" of Products in the Fashion System. Rome: Fondimpresa; 2013. pp. 17

[43] Snyder RL. Fugiive Denim: A Moving Story of People and Pants in a Borderless World. 1st ed. New York: W.W. Norton & Company; 2008

[44] Luiken A. The life cycle of jeans, NVRD-project, chain management jeans. In: ISWA Beacon Conference on Waste Minimisation and Recycling; Wien. 2015

[45] Kenny-Troughton E, Hera L, Haker J. Research report "To what extent is it possible for a denim company or denim department within a company to implement a fully sustainable denim development process?". International Fashion & Management. 2016

[46] Kenny-Troughton E, Hera L, Haker J. Research Report, 500668689. International Fashion & Management. 2016. p. 19

[47] Turunen L, Hayo van der Wert. Life Cycle Analysis of Hemp Textile Yarn—Comparison of Three Hemp Fiber Processing Scenarios and a Flax Scenario. 2016

[48] Addlesperger E. Commodity of the Quarter, Hemp, DePaul University Library, Chicago, Illinois, USA. Journal of Agricultural & Food information. 2015;**16**:196-202. DOI: 10.1080/10496505.2015.1050323

[49] Hayo MG v d W. Life cycle analysis of field production of fibre hemp, the effect of production practices on environmental impacts. Euphytica. 2004;**140**:13-23

[50] She L, Patel MK. Life cycle assessment of polysaccharide materials: A review. Journal of Polymers and Environment. 2008;**16**:154-167. DOI: 10.1007/s10924-008-0092-9

[51] Hayo MG v d W, Turunen L. The environmental impacts of the production of hemp and flax textile yarn. Industrial Crops and Products. 2008;**27**:1-10

[52] Kymalainen H-R, Sioberg A-M. Flax and hemp fibres as raw materials for thermal insulations. Building and Environment. 2008;**43**:1261-1269

[53] Clarke RC. Traditional fiber hemp (cannabis) production, processing, yarn making, and weaving strategies—Functional constraints and regional responses. Part 1. Journal of Natural Fibers. 2010;7(2):118-153. DOI: 10.1080/15440478.2010.482324

[54] Scatol8 official website. Available from: http://scatol8.net/?p=3014

[55] Torres Ana L, Angel GM. Sustainability in Fashion and Textiles: Values, Design, Production and Consumption. Sheffield: Routledge; 2013